T0259435

Wavelets in Image Communication

Series Editor: J. Biemond, Delft University of Technology, The Netherlands

ADVANCES IN IMAGE COMMUNICATION 5

Wavelets in Image Communication

Edited by

M. Barlaud
13S Laboratory URA 1376 CNRS
University of Nice-Sophia Antipolis
Valbonne, France

1994

ELSEVIER
Amsterdam – Lausanne – New York – Oxford – Shannon – Tokyo

ELSEVIER SCIENCE B.V.
Sara Burgerhartstraat 25
P.O. Box 211, 1000 AE Amsterdam, The Netherlands

Library of Congress Cataloging-in-Publication Data

Wavelets in image communication / edited by M. Barlaud.
     p.   cm. — (Advances in image communication ; 5)
    Includes bibliographical references (p.).
    ISBN 0-444-89281-8
    1. Image transmission. 2. Image processing--Digital techniques-
Mathematics. 3. Wavelets (Mathematics) 4. Image compression
I. Barlaud, M. II. Series.
TK5105.2.W38 1994
621.36'7--dc20   94-36820
                                              CIP

ISBN: 0 444 89281 8

This book is printed on acid-free paper.

Transferred to digital printing 2006
Printed and bound by Antony Rowe Ltd, Eastbourne

INTRODUCTION TO THE SERIES
"Advances in Image Communication"

Image Communication is a rapidly evolving multidisciplinary field focussing on the evaluation and development of efficient means for acquisition, storage, transmission, representation and understanding of visual information. Until a few years ago, image communication research was still confined to universities and research laboratories of telecommunication or broadcasting companies. Nowadays, however, this field is also witnessing the strong interest of a large number of industrial companies due to the advent of narrowband and broadband ISDN, digital satellite channels, digital over-the-air transmission and digital storage media. Moreover, personal computers and workstations have become important platforms for multimedia interactive applications that advantageously use a close integration of digital video compression techniques (MPEG), Very Large Scale Integration (VLSI) technology, highly sophisticated network facilities and digital storage media. At the same time, the scope of research of the academic environment on Image Communication has further increased to include model- and knowledge-based image understanding techniques, artificial intelligence, motion analysis, and advanced image and video processing techniques and lead to a diverse area of applications such as: access to image data bases, interactive visual communication, TV and HDTV broadcasting and recording, 3D-TV, graphic arts and communication, image manipulation, etc. The variety of topics on Image communication is so large that no-one can be a specialist in all the topics, and the whole area is beyond the scope of a single volume, while the requirement of up-to-date information is ever increasing.

In 1988, the European Association for Signal Processing EURASIP together with Joel Claypool & Ir. Hans van der Nat, at that time Publishing Editors at Elsevier Science Publishers, conceived several projects to meet this need for information. First of all a new EURASIP journal, "Signal Processing: Image Communication", was launched in June 1989 under the inspired Editorship of Dr. Leonardo Chiariglione. So far, the journal has been a major success not in the least due to the many special issues devoted to timely aspects in Image Communication, such as low/medium/high bit rate video coding, all digital HDTV, 3D-TV, etc. It was further decided to publish a book series in the field, an idea enthusiastically supported by Dr. Chiariglione. Mr. van der Nat approached the undersigned to edit this series.

It was agreed that the book series should be aimed to serve as a comprehensive reference work for those already active in the area of Image Communication. Each volume author or editor was asked to write or compile a state-of-the-art book in his area of expertise, and containing information until now scattered in many journals and proceedings. The book series therefore should help Image Communication specialists to get a better understanding of the important issues in neighbouring areas by reading particular volumes. At the same time, it should give newcomers to the field a foothold for doing research in the Image Communication area. In order to produce a quality book series, it was necessary to ask authorities well known in their respective fields to serve as volume editors, who would in turn attract outstanding contributors. It was a great pleasure to me that ultimately we were able to attract such an excellent team of editors and authors.

The Series Editor wishes to thank all of the volume editors and authors for the time and effort they put into the book series. He is also grateful to Ir. Hans van der Nat and Drs. Mark Eligh of Elsevier Science Publishers for their continuing effort to bring the book series from the initial planning stage to final publication.

Jan Biemond
Delft University of Technology
Delft, The Netherlands
1993

Future titles planned for the series "Advances in Image Communication":

– Subband Coding of Images	T.A. Ramstad, S.O. Aase
– HDTV Signal Processing	R. Schäfer, G. Schamel
– Magnetic Recording	M. Breeuwer, P.H.N. de With
– Image Deblurring; Motion Compensated Filtering	A.K. Katsaggelos, N. Galatsanos
– Colour Image Processing	P.E. Trahanias, A.N. Ventsanopoulos

PREFACE

This book is concerned with the definition, study and use of the wavelet transform in communications for two-dimensional digital signals.

This transform is used for signal reorganization before compression: We present and discuss the trade-off between these two steps and the whole compression process. The transform must be simultaneously adapted to the image signal, to the quantization, and to the coding process. Our goal is to show that the proposed transform is well adapted for image compression.

Signal compression needs powerful tools for information "reorganization"; such tools include predictive techniques for voice coding or DCT-based coding for images or video that are now standards like JPEG and MPEG.

The cosine functions used in the DCT are nearly optimal for stationary signals, but images are rarely stationary: DCT-based algorithms do not tolerate high compression ratios. The wavelet transform, due to the good localization in space and frequency domain of the basis functions, can handle non-stationary signals, hence greater compression ratios can be obtained.

We then briefly recall the history of wavelets, but from the image processing point of view. Many researchers have worked on this theme but we cannot quote all of them.

The wavelet transform first appeared as an "ondelette" for analyzing seismic responses (J. Morlet 83). Mathematical models of this first wavelet were then developed (Grossman 84), leading to the "orthogonal wavelet basis" (Y. Meyer 86). Mallat (89) subsequently provided a fast algorithm for this transform, applied to images. I. Daubechies (88) constructed orthonormal bases of compactly supported wavelets, showing the link between wavelets and FIR filters.

Of key interest for image coding is the development of the biorthogonal wavelet theory, enabling the definition of different filter banks for analysis and synthesis. It converges with Smith and Barnwell's (86) previous proposition of exact reconstruction filters for coding purposes, in an other approach. The extension to two dimensions is often realized in a separable way, but the non-separable design of the biorthogonal wavelet transform for image processing permits a finer decomposition in scale and a better isotropy. Many other wavelet-like transforms have been proposed: Wave-packets by Wickerhauser (91), lapped transform by E. Malvar (89).

As a matter of fact, the four advantages of the wavelet transform for image compression are the notion of multiresolution, the link with digital filters, the fast algorithm, and the linear phase. Multiresolution offers the means to optimize bit allocation, but also to process in a pyramidal manner like Burt and Adelson (83). Regularity, another property of the wavelet function, can be used to analyze and design filter banks.

This book is divided into five chapters which present the theory of wavelets applied to images, then applications of compression of still images and sequences.

The first chapter introduces biorthogonal bases of compactly supported wavelets. This generalization of orthonormal wavelet theory, allows the use of linear phase filters. Two non separable 2D cases are presented.

In the second chapter, a non rectangular wavelet representation of 2D signals is developed. The properties usually used with wavelets are discussed: phase linearity, regularity.

Chapter 3 is composed of three parts: first a description of commonly used biorthogonal wavelets; second, a presentation of vector quantization algorithms. Lattice vector quantization is described, followed by a discussion of the bit allocation procedure, and finally, experimental results are given.

Chapter 4 deals with a region-based discrete wavelet transform for image coding.

Region based coding methods divide the image into regions and code the segmentation information and the texture separately. A region based discrete wavelet transform is presented to code texture.

The last chapter investigates the transmission of image sequences. Wavelet transforms and motion estimation are detailed in a multiconstraint approach of image sequence coding.

CONTENTS

Chapter 3
WAVELET TRANSFORM AND IMAGE CODING
M. Antonini, T. Gaidon, P. Mathieu and M. Barlaud

Chapter 4
A REGION-BASED DISCRETE WAVELET TRANSFORM FOR IMAGE CODING

H.J. Barnard, J.H. Weber and J. Biemond

Chapter 1

Biorthogonal Wavelets and Dual Filters

ALBERT COHEN

CEREMADE,
Université de Paris IX-Dauphine,
Place du Maréchal de Lattre de Tassigny
75775 Paris Cedex 16

Abstract

In this chapter, we introduce biorthogonal bases of compactly supported wavelets which are useful for signal and image processing. This construction is based on subband coding schemes with perfect reconstruction and finite impulse response filters. It constitutes an important generalization of orthonormal wavelets since it allow to use linear phase filters. We give several examples in one and two dimensions.

I. Introduction.

Biorthogonal wavelet bases consist in a pair of Riesz bases generated from two single compactly supported functions by mean of dilations and translations

(1.1)
$$\begin{cases} \psi_k^j(x) = 2^{-j/2}\,\psi(2^{-j}x - k) & j \in \mathbf{Z}, \ k \in \mathbf{Z} \\ \tilde{\psi}_k^j(x) = 2^{-j/2}\,\tilde{\psi}(2^{-j}x - k) & j \in \mathbf{Z}, \ k \in \mathbf{Z} \end{cases}$$

Any function f can be expressed uniquely with the two stable decompositions

(1.2)
$$f = \sum_{j,k \in \mathbf{Z}} <f|\tilde{\psi}_k^j> \psi_k^j = \sum_{j,k \in \mathbf{Z}} <f|\psi_k^j> \tilde{\psi}_k^j$$

Remark that it is false in general that a Riesz basis of the type $\{\psi_k^j\}_{j,k \in \mathbf{Z}}$ has a dual basis of the same type.

The goal of this construction is to mimic, in a more general setting (more convenient for signal processing) the construction of orthonormal bases of compactly supported wavelets developed in [Dau1] that we briefly recall here in three steps :

· Orthonormal wavelets are associated with a scaling function φ which defines a multiresolution analysis, i.e. a ladder of embedded approximation subspaces of $L^2(\mathbf{R})$

(1.3)
$$\{0\} \rightarrow \ldots V_1 \subset V_0 \subset V_{-1} \ldots \rightarrow L^2(\mathbf{R})$$

such that $\{\varphi_k^j\}_{k \in \mathbf{Z}} = \{2^{-j/2}\,\varphi(2^{-j}x - k)\}_{k \in \mathbf{Z}}$ is an orthonormal basis of V_j. The wavelets are built to characterize the missing details between two adjacent level of approximation. More precisely, the set $\{\psi_k^j\}_{k \in \mathbf{Z}} = \{2^{-j/2}\,\psi(2^{-j}x - k)\}_{k \in \mathbf{Z}}$ is an orthonormal basis for the orthogonal complement of V_j into V_{j+1}.

· The construction of φ and ψ is based on a trigonometric polynomial $m_0(\omega)$ such that $m_0(0) = 1$ and

(1.4)
$$|m_0(\omega)|^2 + |m_0(\omega + \pi)|^2 = 1 .$$

The functions φ and ψ are then defined by

(1.5)
$$\hat{\varphi}(\omega) = \prod_{k=1}^{+\infty} m_0(2^{-k}\omega)$$

$$(1.6) \qquad \hat{\psi}(\omega) = m_1\left(\frac{\omega}{2}\right)\hat{\varphi}\left(\frac{\omega}{2}\right) = e^{-i\frac{\omega}{2}}\,\overline{m_0\left(\frac{\omega}{2}+\pi\right)}\,\hat{\varphi}\left(\frac{\omega}{2}\right).$$

· In the Fast Wavelet Transform algorithm (FWT), m_0 and m_1 can be viewed as the transfer functions of a low pass and a high pass filter that split the discrete signal into two channels. The new signals are decimated of one sample out of two and the process is iterated on the low pass channel. The final result is a coarse approximation of the original signal and a succession of details at each intermediate scale. Perfect reconstruction is performed by the same filters which are used to interpolate and refine the decimated channels.

Unfortunately, these filters known as conjugate quadrature filters (CQF see [SB]) have some disadvantages for practical design and applications :

· They cannot be both FIR and linear phase (i.e. with real and symmetrical coefficients) except the Haar filter which does not lead to a continuous φ and ψ.
· They are solutions of the quadratic equation (1.4) and their coefficients are usually algebraic numbers with no simple expression.
· Their design uses the Fejer-Riesz factorization lemma (see [Dau1]) which does not generalize in the multidimensional framework.
· When these filter have finite impulse response, the spaces V_j have no simple direct definition. In particular, they cannot be spline functions except in the Haar case.

For all these reasons, we considered a larger class of filters leading to biorthogonal bases. The idea is to allow the analyzing and reconstructing filter to be different. We first describe this construction in a theoretical way.

II. The construction of biorthogonal wavelets.

The starting point to biorthogonal wavelets is a two channel subband coding scheme with perfect reconstruction. It can be represented as in Figure 1 : the original discrete signal is divided in two channels by the action of two discrete filters represented by their transfer function $\overline{\tilde{m}_0(\omega)}$ (low pass) and $\overline{\tilde{m}_1(\omega)}$ (high pass), followed by a decimation of one sample out of two. The reconstruction is made by inserting zeros in these two subsampled channels and interpolating with the filters $m_0(\omega)$ (low pass) and $m_1(\omega)$ (high pass).

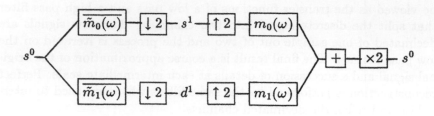

<u>Figure 1</u>

Perfect reconstruction subband coding scheme

($\downarrow$ stands for downsampling, $\uparrow$ for upsampling)

Perfect reconstruction without aliasing will be achieved for any signal s^0 if and only if the filters satisfy the equations

(2.1)
$$
\begin{cases}
\overline{\tilde{m}_0(\omega)}\, m_0(\omega) + \overline{\tilde{m}_1(\omega)}\, m_1(\omega) = 1 \\
\overline{\tilde{m}_0(\omega + \pi)}\, m_0(\omega) + \overline{\tilde{m}_1(\omega + \pi)}\, m_1(\omega) = 0
\end{cases}
$$

If we want FIR filters, the determinant of this system should be a monomial. Up to a shift on the coefficients, we choose $\Delta = -e^{i\omega}$ which leads to

(2.2)
$$ m_1(\omega) = e^{-i\omega}\, \overline{\tilde{m}_0(\omega + \pi)} $$

(2.3)
$$ \tilde{m}_1(\omega) = e^{-i\omega}\, \overline{m_0(\omega + \pi)} $$

and

(2.4)
$$ \overline{m_0(\omega)}\, \tilde{m}_0(\omega) + \overline{m_0(\omega + \pi)}\, \tilde{m}_0(\omega + \pi) = 1 \,. $$

Remark that when $m_0 = \tilde{m}_0$, $m_1 = \tilde{m}_1$ we find the CQF-orthonormal case presented in the introduction. We also assume that these filters satisfy

(2.5)
$$ m_0(0) = \tilde{m}_0(0) = 1 \,, $$

and

(2.6) $$m_0(\pi) = \tilde{m}_0(\pi) = 0 .$$

We can define $\varphi, \tilde{\varphi}, \psi$ and $\tilde{\psi}$, first as compactly supported distributions, by the formulas

(2.7) $$\hat{\varphi}(\omega) = \prod_{k=1}^{+\infty} m_0(2^{-k}\omega) \quad \text{and} \quad \hat{\psi}(\omega) = m_1\left(\frac{\omega}{2}\right)\hat{\varphi}\left(\frac{\omega}{2}\right) ,$$

(2.8) $$\widehat{\tilde{\varphi}}(\omega) = \prod_{k=1}^{+\infty} \tilde{m}_0(2^{-k}\omega) \quad \text{and} \quad \widehat{\tilde{\psi}}(\omega) = \tilde{m}_1\left(\frac{\omega}{2}\right)\widehat{\tilde{\varphi}}\left(\frac{\omega}{2}\right) .$$

The main problem is the convergence of the infinite products. It can be checked (see [Me], [Co1]) that in the orthonormal case these functions are always in $L^2(\mathbb{R})$. This is not true anymore in the biorthogonal setting and we have to choose more carefully the filters m_0 and $\tilde{m}_0$.

Let us describe, with sketches of the proof, three results contained in [CDF] or [Co2] that lead to biorthogonal wavelet bases.

· The first result (Theorem 2.1 in [Co2]) states that if $\varphi, \tilde{\varphi}, \psi$ and $\tilde{\psi}$ are in $L^2(\mathbb{R})$, then for all $J > 0$ and all f in $L^2(\mathbb{R})$,
(2.9)

$$\sum_{k\in\mathbb{Z}} <f|\tilde{\varphi}_k^{-J}> \varphi_k^{-J} = \sum_{k\in\mathbb{Z}} <f|\tilde{\varphi}_k^{J}> \varphi_k^{J} + \sum_{j=1-J}^{J}\sum_{k\in\mathbb{Z}} <f|\tilde{\psi}_k^{j}> \psi_k^{j}$$

and by letting J go to $+\infty$,

(2.10) $$f = \lim_{J\to+\infty} \sum_{j=-J}^{J}\sum_{k\in\mathbb{Z}} <f|\tilde{\psi}_k^{j}> \psi_k^{j} .$$

These formulas are of course still valid if we exchange (φ, ψ) and $(\tilde{\varphi}, \tilde{\psi})$. (2.9) is obtained by using the perfect reconstruction properties of the filters and (2.10) uses the fact that φ and $\tilde{\varphi}$ have smoothing properties.

This does not mean yet that ψ_k^{j} and $\tilde{\psi}_k^{j}$ are biorthogonal bases since the decomposition in (2.10) could be redundant and not stable.

· The second result (Lemma 2.6 in [Co2], 3.6 and 3.7 in [CDF]) tells us that the linear independence and biorthogonality for the ψ_k^j and $\tilde{\psi}_k^j$ holds if and only if the approximants φ_n and $\tilde{\varphi}_n$ defined by

$$(2.11) \qquad \hat{\varphi}_n(\omega) \;=\; \prod_{k=1}^{n} m_0(2^{-k}\omega)\,\chi_{[-2^n\pi,\,2^n\pi]}(\omega)$$

$$(2.12) \qquad \hat{\tilde{\varphi}}_n(\omega) \;=\; \prod_{k=1}^{n} \tilde{m}_0(2^{-k}\omega)\,\chi_{[-2^n\pi,\,2^n\pi]}(\omega)$$

converge to φ and $\tilde{\varphi}$ in $L^2(\mathbb{R})$.

One proves indeed by recursion that for all $n > 0$,

$$(2.13) \qquad < \varphi_n(x-k)\,|\,\tilde{\varphi}_n(x-l) > \;=\; \delta_{k,l}$$

and at the limit

$$(2.14) \qquad < \varphi(x-k)\,|\,\tilde{\varphi}(x-l) > \;=\; \delta_{k,l}\;.$$

This is equivalent to

$$(2.15) \qquad < \psi_k^j\,|\,\tilde{\psi}_{k'}^{j'} > \;=\; \delta_{j,j'}\,\delta_{k,k'}$$

by scaling arguments and using the perfect reconstruction identities.

The duality relations (2.15) imply the linear independence and the decomposition in (2.10) is thus unique. We still have to check the stability, i.e. that ψ_k^j and $\tilde{\psi}_k^j$ are Riesz bases.

· The third result (proposition 4.9 and theorem 3.2 in [CDF] or theorem 3.1 in [Co2]) gives a sufficient condition for stability : If we use the factorized forms

$$(2.16) \qquad m_0(\omega) \;=\; \left(\frac{1+e^{i\omega}}{2}\right)^N p(\omega)$$

$$(2.17) \qquad \tilde{m}_0(\omega) \;=\; \left(\frac{1+e^{i\omega}}{2}\right)^{\tilde{N}} \tilde{p}(\omega)$$

and assume that forme some $k, \tilde{k} > 0$

$$(2.18) \qquad B_k^{'} = \max_{\omega \in \mathbb{R}} |p(\omega)p(2\omega)\dots p(2^{k-1}\omega)| < 2^{N-\frac{1}{2}}$$

$$(2.19) \qquad \tilde{B}_{\tilde{k}} = \max_{\omega \in \mathbb{R}} |\tilde{p}(\omega)\tilde{p}(2\omega)\dots \tilde{p}(2^{\tilde{k}-1}\omega)| < 2^{\tilde{N}-\frac{1}{2}}$$

then, the hypotheses of the two previous results are satisfied and moreover, there exist two constants $A, \tilde{A} > 0$, such that for all f in $L^2(\mathbb{R})$

$$(2.20) \qquad \tilde{A}^{-1} \|f\|^2 \leq \sum_{j,k \in \mathbb{Z}} |< f|\psi_k^j >|^2 \leq A \|f\|^2$$

$$(2.21) \qquad A^{-1} \|f\|^2 \leq \sum_{j,k \in \mathbb{Z}} |< f|\tilde{\psi}_k^j >|^2 \leq \tilde{A} \|f\|^2$$

which means that ψ_k^j and $\tilde{\psi}_k^j$ are Riesz bases. Note that these inequalities ensure the stability of the decomposition-reconstruction algorithm.

The estimates (2.18) and (2.19) easily lead to the boundedness of $(1 + |\omega|)^{\varepsilon+\frac{1}{2}} (|\hat{\varphi}(\omega)| + |\hat{\tilde{\varphi}}(\omega)|)$ for some $\varepsilon > 0$.

The upper bounds in (2.20) and (2.21) are obtained by using this boundedness with a Poisson summation formula in the estimates of the l^2 norm of the coordinates. The lower bounds are derived by computing $\|f\|^2$ from (2.10) and using the Schwarz inequality.

One can also use this estimate on spectral decay of φ and $\tilde{\varphi}$ to derive a lower bound for the regularity of these functions. More precisely if for some $k, \tilde{k} > 0$ and $\alpha, \tilde{\alpha} > 0$,

$$(2.22) \qquad B_k < 2^{N-1-\alpha} \quad \text{and} \quad \tilde{B}_{\tilde{k}} < 2^{\tilde{N}-1-\tilde{\alpha}},$$

then $|\hat{\varphi}(\omega)|(1 + |\omega|)^{1+\alpha+\varepsilon}$ and $|\hat{\tilde{\varphi}}(\omega)|(1 + |\omega|)^{1+\tilde{\alpha}+\varepsilon}$ are uniformly bounded for some $\varepsilon > 0$ and thus $\varphi \in \mathcal{C}^\alpha$ and $\tilde{\varphi} \in \mathcal{C}^{\tilde{\alpha}}$.

Sharper conditions for regularity are given in [CD1] and [DL] and necessary and sufficient conditions on m_0 and $\tilde{m}_0$ to generate biorthogonal bases are given in [CD2]. We shall not go any further into the theoretical aspects of biorthogonal wavelet bases since our purpose is mainly to present a large number of examples that are useful in signal and image processing.

We start with the one-dimensional examples.

III. Examples in $1D$.

Recall the duality relation on the filters expressing the property of perfect reconstruction

$$(3.1) \qquad \overline{m_0(\omega)}\,\tilde{m}_0(\omega) + \overline{m_0(\omega + \pi)}\,\tilde{m}_0(\omega + \pi) \; = \; 1$$

Such filters are relatively easy to design. In particular if we choose to fix $m_0(\omega)$, finding a $\tilde{m}_0(\omega)$ that solves (3.1) is only a Bezout problem: solution will exist in infinite number if and only if the extension of $m_0(\omega) = m_0(e^{i\omega})$ from the unit circle to the complex plane has no zeros of the type $\{z, -z\}$, with $z \neq 0$. How can we choose $\tilde{m}_0$ among the different solutions ? It of course depends on the application, but for many problems such as compression, coding or approximation, the regularity of the scaling function and the wavelet is important essentially for the reconstruction. This means that once m_0 is chosen such that φ and ψ are regular, we can look for the smallest filter $\tilde{m}_0$ (i.e. with the smallest number of coefficients) such that (3.1) is satisfied and that $\tilde{\varphi}$ and $\tilde{\psi}$ are in $L^2(\mathbb{R})$ (in order to have biorthogonal wavelet bases and stable algorithm, according to the results of the previous section).

III.A - Biorthogonal splines.

If we want to build spline wavelets of degree N, i.e. piecewise polynomial of degree N and globally $\mathcal{C}^{N-1}$, then a natural choice for the scaling function is the N-th degree box spline defined by

$$(3.2) \qquad \varphi^N(x) \; = \; (*)^{N+1}\,\chi_{[0,1]}$$

Clearly, except for the Haar case ($N = 0$), the translates of ϕ are not orthogonal. It is possible to build a new function which achieve this property by the formula

$$(3.3) \qquad \hat{\varphi}_O^N \; = \; \hat{\varphi}^N(\omega)\left(\sum_{k \in \mathbb{Z}} |\hat{\varphi}^N(\omega + 2k\pi)|^2\right)^{-1/2}$$

but unfortunately the new spline function φ_O^N is not compactly supported and so will be the associated wavelet. As a consequence, the

CQF filters have not a finite impulse response which is a major disadvantage for the implementation.

In the biorthogonal setting, we can keep φ^N as a scaling function. The generating filter is given by

$$(3.4) \qquad m_0^N(\omega) = \left(\frac{1+e^{-i\omega}}{2}\right)^{N+1}.$$

One easily checks that $\hat{\varphi}^N(\omega) = \prod_{k=1}^{+\infty} m_0^N(2^{-k}\omega)$. To find a dual filter, we use the polynomial

$$(3.5) \qquad P_L(y) = \sum_{j=0}^{L-1} \binom{L-1+j}{j} y^j$$

which solves the Bezout problem

$$(3.6) \qquad (1-y)^L P_L(y) + y^L P_L(1-y) = 1.$$

By the change of variable $y = \sin^2\left(\frac{\omega}{2}\right)$, we obtain

$$(3.7) \qquad \left[\cos\left(\frac{\omega}{2}\right)\right]^{2L} P_L\left(\sin^2\left(\frac{\omega}{2}\right)\right) + \left[\sin\left(\frac{\omega}{2}\right)\right]^{2L} P_L\left(\cos^2\left(\frac{\omega}{2}\right)\right) = 1$$

and by a shift

$$(3.8) \qquad \begin{aligned} &\left(\frac{1+e^{i\omega}}{2}\right)^{2L}\left[e^{-iL\omega} P_L\left(\sin^2\left(\frac{\omega}{2}\right)\right)\right] + \\ &\left(\frac{1-e^{i\omega}}{2}\right)^{2L}\left[(-1)^L e^{-iL\omega} P_L\left(\cos^2\left(\frac{\omega}{2}\right)\right)\right] = 1 \end{aligned}$$

If $N+1 \leq 2L$, we can thus take as a dual filter of $m_0^N(\omega)$

$$(3.9) \qquad \tilde{m}_0^{N,L}(\omega) = \left(\frac{1+e^{i\omega}}{2}\right)^{2L-N-1} P_L\left(\sin^2\left(\frac{\omega}{2}\right)\right) e^{-iL\omega}.$$

For a fixed value of N, we need to choose a good value for L. It turns out that, except for the Haar case ($N = 0$), the smallest value of L such that $2L \geq N+1$ does not lead to a square-integrable $\tilde{\varphi}$ and we need to take a larger value.

For $N = 1$, the smallest possible value is $L_{\min} = 2$, i.e.

$$(3.10) \qquad \tilde{m}_0^{1,2} = \left(\frac{1 + e^{i\omega}}{2}\right) e^{-2i\omega} \left(1 + 2\sin^2\left(\frac{\omega}{2}\right)\right).$$

For $N = 2$, $L_{\min} = 2$ and for $N = 3$, $L_{\min} = 4$.

Figure 2 and 3 illustrate the linear and the quadratic case ($N = 1$ and 2). We show the aspects of $\varphi, \tilde{\varphi}, \psi, \tilde{\psi}$ for the choices $L_{\min}$ and $L_{\min} + 1$. The regularity of $\tilde{\varphi}$ and $\tilde{\psi}$ increases with L.

III.B - The Burt-Adelson wavelets.

The filters used by Burt and Adelson for image processing are a one-parameter family given by

$$(3.11) \qquad m_0^a(\omega) = \cos^2\left(\frac{\omega}{2}\right)[4a - 1 + (2 - 4a)\cos\omega]$$

or $m_0^a(\omega) = h_0 + h_1(e^{i\omega} + e^{-i\omega}) + h_2(e^{2i\omega} + e^{-2i\omega})$ with

$$(3.12) \qquad h_0 = a \quad, \quad h_1 = \frac{1}{4} \quad \text{and} \quad h_2 = \frac{1}{4} - \frac{a}{2}.$$

The technique used by Burt and Adelson to obtain the details consists in substracting to the original signal its approximation obtained by the action of $m_0(\omega)$ followed by a decimation and interpolation with the same filter. This leads to some redundancy in the high pass channel which can be avoided by using a dual filter of m_0^a to reconstruct the signal. Once again, we need to solve (3.1). If we impose that $\tilde{m}_0^a(\pi) = 0$ the minimal choice is a seven taps filter given by

$$(3.13)$$
$$\tilde{m}_0^a(\omega) = \frac{\cos^2\left(\frac{\omega}{2}\right)}{4a - 1}\left[8a^2 - 10a + 5(3-4a)^2\cos\omega + (3-4a)(1-2a)\cos 2\omega\right]$$

or

$$(3.14) \qquad \begin{cases} \tilde{h}_0 = \dfrac{1 + 4a}{4(4a - 1)} \quad, \quad \tilde{h}_1 = \dfrac{-8a^2 + 18a - 5}{8(4a - 1)}, \\[3mm] \tilde{h}_2 = \dfrac{4a - 3}{8(4a - 1)} \quad, \quad \tilde{h}_3 = \dfrac{(1 - 2a)(3 - 4a)}{8(4a - 1)}. \end{cases}$$

Remark that this filter is not defined for $a = \frac{1}{4}$. This is due to the fact that in this case $m_0^{1/4}(\omega) = \cos^2\left(\frac{\omega}{2}\right)\cos\omega$ which does not satisfy the hypothesis of non degeneracy for the Bezout problem (3.1).

On Figure 4, we plot the functions $\varphi, \tilde{\varphi}, \psi$ and $\tilde{\psi}$ in the case $a = 0.6$. This case is particularly interesting because the coefficients of m_0^a and $\tilde{m}_0^a$ are very close. As a result, the analyzing functions are similar to the synthesis functions. This means that ψ_k^j and $\tilde{\psi}_k^j$ are "almost" orthonormal bases and that their conditioning numbers A and $\tilde{A}$ are very close to 1.

III.C - Optimizing the dual filter design.

The flexibility of dual filters allow to optimize their design. In a joint work with Jim Johnston [JC], we investigate the possibilities of frequency localization offered by these filters, for a given number of taps, with an optimization algorithm. The requirement is here that $m_0(\omega)$ and $\tilde{m}_0(\omega)$ must be as close as possible to the ideal filter corresponding to the transfer function $\chi_{[-\pi/2,\pi/2]}(\omega)$. The principle is the following :

1) Start from a given filter $m_0(\omega)$.
2) Derive the smallest dual filter $\tilde{m}_0(\omega)$.
3) Estimate by a certain criterion their frequency localization.
4) Perturbate $m_0(\omega)$ and iterate the process to improve this localization.

This work is not published yet but the first results seems very encouraging for the signal processing applications.

All these examples illustrate the advantages of biorthogonal wavelets compared to orthonormal wavelets. Their design is simplier and more flexible. These qualities will be confirmed in the bidimensional case that we now investigate.

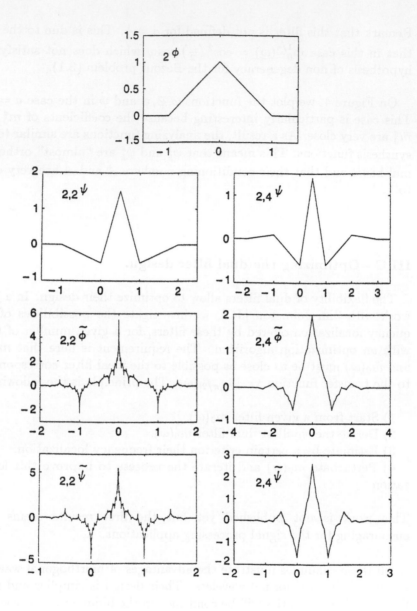

<u>Figure 2</u>

Graphs of $\varphi, \tilde{\varphi}, \psi$ and $\tilde{\psi}$ in the linear case ($N = 1$, $L = 2$ and 3)

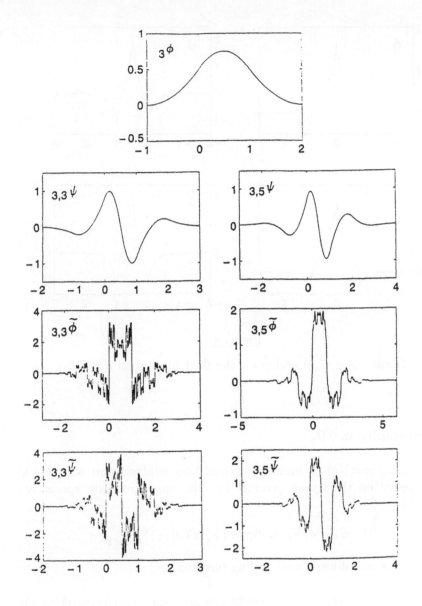

Figure 3

Graphs of $\varphi, \tilde{\varphi}, \psi$ and $\tilde{\psi}$ in the quadratic case ($N = 2$, $L = 2$ and 3)

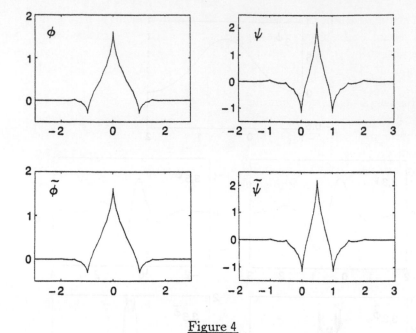

Figure 4

Graphs of $\varphi, \tilde{\varphi}, \psi$ and $\tilde{\psi}$ for the Burt filter with $a = 0.6$

IV. Examples in $2D$.

The usual method to build multiresolution analysis and wavelets in $2D$ is based on the tensor product i.e. the approximation spaces are defined by

$$(4.1) \qquad \mathcal{V}_j \ = \ V_j \otimes V_j \ = \ \text{Span} \left\{ \varphi_k^j(x)\,\varphi_l^j(y) \right\}_{(k,l) \in \mathbf{Z}^2}$$

where φ is a one dimensional scaling function.

As a consequence, three wavelets are necessary to characterize the details

$$\psi_a(x,y) \ = \ \psi(x)\,\varphi(y) \quad , \quad \psi_b(x,y) \ = \ \varphi(x)\,\psi(y)$$
$$(4.2) \qquad \text{and} \quad \psi_c(x,y) \ = \ \psi(x)\,\psi(y)$$

which represent in image processing respectively the horizontal, vertical and "corner" edges (see [**Ma**]).

In the algorithm, this means that the one dimensional filters are used in a separable fashion : the signal is filtered and decimated along the rows and the resulting channels are again filtered and decimated along the column. The reconstruction is performed in the same separable way.

This technique is very simple but also clearly restrictive among all the other possibilities. Moreover it leads to a non isotropical analysis since the horizontal and vertical directions have a particular importance. To circumvent this drawback we have to use non separable filters. Such filters are very difficult to design in the orthonormal case. Indeed the CQF equation involves the square modulus $|m_0|^2$ of the transfert function. In $1D$, it is possible (see [Dau1]) to design $|m_0|^2$ as a positive trigonometric polynomial and derive m_0 by the Fejer-Riesz factorization lemma. Unfortunately this lemma does not generalize to more than $1D$. We shall see that biorthogonal wavelet bases constitute a more appropriate framework to design non separable filter.

IV.A - The quincunx sublattice construction.

This technique consists in replacing the scaling factor 2 by a scaling matrix with integer entries such as

$$(4.2) \qquad R = \begin{pmatrix} 1 & 1 \\ -1 & 1 \end{pmatrix} \quad \text{or} \quad S = \begin{pmatrix} 1 & 1 \\ 1 & -1 \end{pmatrix}$$

These matrices have determinant 2 or -2 and maps $\mathbf{Z}^2$ into the "quincunx sublattice" Q (Figure 5) which is obtained by removing one diagonal out of two.

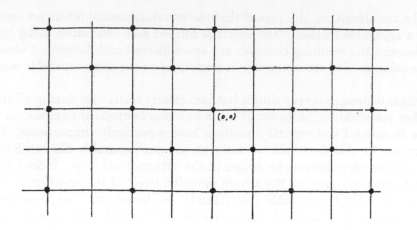

Figure 5

The quincunx sublattice Q

In an algorithmic point of view, this means that we only use two channels (instead of four in the separable case). Clearly the horizontal, vertical and diagonal direction play the same role here since they are exchanged under the action of R or S. To take profit of this property, it is necessary to build symmmetrical filters around these four directions.

In this case the perfect reconstruction equation is

$$(4.3) \qquad \overline{m}_0 \, \tilde{m}_0(\omega_1, \omega_2) + \overline{m}_0 \, \tilde{m}_0(\omega_1 + \pi, \omega_2 + \pi) = 1 \, .$$

A possibility to design this type of filters is to use the Mc Clellan transform i.e. start from a pair of one dimensional symmetrical dual filters

$$(4.4) \qquad m_0(\omega) = p_0(\cos \omega) \quad \text{and} \quad \tilde{m}_0(\omega) = \tilde{p}_0(\cos \omega)$$

and define

$$(4.5) \quad \begin{cases} m_0(\omega_1, \omega_2) = p_0 \left(\dfrac{\cos \omega_1 + \cos \omega_2}{2} \right) \\ \text{and} \\ \tilde{m}_0(\omega_1, \omega_2) = \tilde{p}_0 \left(\dfrac{\cos \omega_1 + \cos \omega_2}{2} \right) \end{cases}$$

Clearly these filters have the good symmetries. Moreover, we have shown in [CD1] that for the one dimensional family of splines wavelet described in section III.A, the regularity of the $2D$ wavelets increases linearly with the regularity of the $1D$ wavelet so that it can be arbitrarily high.

An open problem is to know if this is true for any symmetrical wavelet i.e. if the Mc Clellan transform always leads to smooth $2D$ wavelets provided that the associated $1D$ wavelets have a certain regularity.

On Figure 6 and 7 we show the aspects of $\varphi, \tilde{\varphi}, \psi$ and $\tilde{\psi}$ corresponding to the Mc Clellan transform in the case of linear splines. They are associated to the $1D$ filters

$$(4.6) \qquad m_0^1(\omega) = \left|\frac{1+e^{i\omega}}{2}\right|^2 = \cos^2\left(\frac{\omega}{2}\right)$$

$$(4.7) \qquad \tilde{m}_0^{1,3}(\omega) = \cos^2\left(\frac{\omega}{2}\right)\left[1 + 3\,\sin^2\left(\frac{\omega}{2}\right) + 6\,\sin^4\left(\frac{\omega}{2}\right)\right].$$

The Hölder exponent for φ and ψ is $\cdot 61$ (see [CD1]).

These functions are not exactly radial but they are much more isotropic than what can be obtain with the tensor product. The wavelets ψ and $\tilde{\psi}$ for example are close to the laplacian of a radial smoothing function. This means that all the edges, whatever their orientation may be, will be matched with the same intensity in the detail channel at a given scale.

These wavelets have been used by M. Barlaud for image coding and compression and the results seem better than in the separable case where some edge artefacts can be caused by the lack of isotropy.

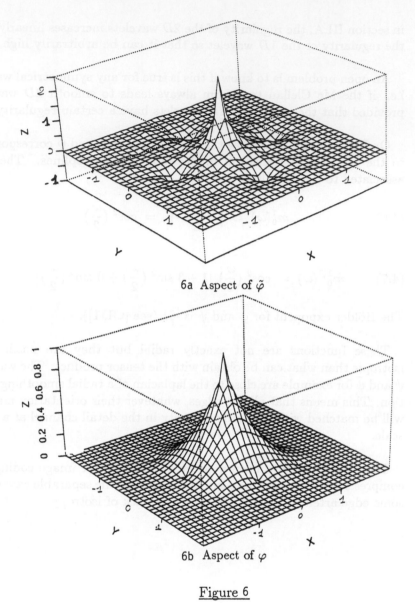

6a Aspect of $\tilde{\varphi}$

6b Aspect of φ

Figure 6

The scaling functions $\tilde{\varphi}$ and φ (Quincunx)

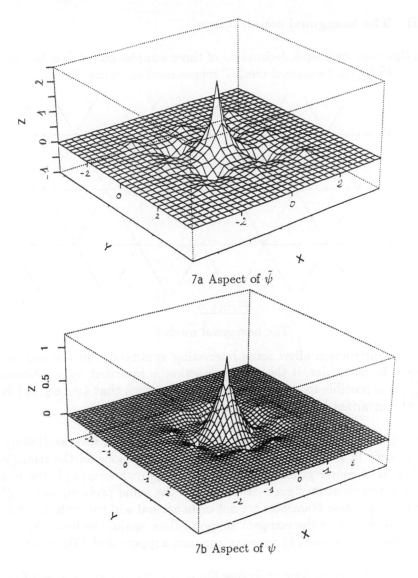

7a Aspect of $\tilde{\psi}$

7b Aspect of ψ

Figure 7

The wavelets ψ and $\tilde{\psi}$ (Quincunx)

IV.B - The hexagonal construction.

In this case, we keep a decimation of three samples out of four but we replace $\mathbf{Z}^2$ by the hexagonal mesh Γ represented on figure 8.

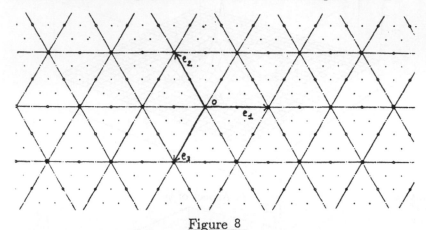

Figure 8

The hexagonal mesh Γ

This construction allow some interesting symmetries in the wavelet design. In particular, if the scaling function is invariant by a rotation of $\frac{2\pi}{3}$, it is possible to build the three wavelets such that $\{\psi_1, \psi_2, \psi_3\}$ is globally invariant by the same rotation.

A typical example is the linear spline bidimensional wavelet ([**Me**]). In this case V_0 is the space of piecewise affine functions on the triangle of Γ. It is of course generated by $\{\varphi(x - \gamma)\}_{\gamma \in \Gamma}$ where $\varphi(x)$ is the hat function represent on figure 10a. Clearly the same problem as in $1D$ occurs here : these translates are not orthonormal and the orthonormalization process kills the compact support. Here again, the biorthogonal framework can be used to preserve compact support and FIR filters.

We have to design four analyzing filters $\tilde{m}_0, \tilde{m}_1, \tilde{m}_2$ and $\tilde{m}_3$ and four reconstructing filters m_0, m_1, m_2 and m_3. To express the symmetries we use the non-independent variables in the frequency plane

$$(4.8) \quad \omega_1 = <\omega|e_1> \quad , \quad \omega_2 = <\omega|e_2> \quad , \quad \omega_3 = <\omega|e_3> = -\omega_1 - \omega_2$$

and write

$$(4.9) \quad m_0(\omega_1, \omega_2, \omega_3) = m_0(\omega_2, \omega_3, \omega_1) = m_0(\omega_3, \omega_1, \omega_2)$$

and

(4.10) $$m_2(\omega_1, \omega_2, \omega_3) = m_1(\omega_2, \omega_3, \omega_1)$$

(4.11) $$m_3(\omega_1, \omega_2, \omega_3) = m_1(\omega_3, \omega_1, \omega_2)$$

and similarly for the analyzing filters.

The exact reconstruction condition leads to the system

(4.12)
$$\begin{cases} \sum_{i=0}^{3} \overline{m_i(\omega_1, \omega_2, \omega_3)}\, \tilde{m}_i(\omega_1, \omega_2, \omega_3) = 1 \\[2mm] \sum_{i=0}^{3} \overline{m_i(\omega_1, \omega_2, \omega_3)}\, \tilde{m}_i(\omega_1 + \pi, \omega_2 + \pi, \omega_3) = 0 \\[2mm] \sum_{i=0}^{3} \overline{m_i(\omega_1, \omega_2, \omega_3)}\, \tilde{m}_i(\omega_1 + \pi, \omega_2, \omega_3 + \pi) = 0 \\[2mm] \sum_{i=0}^{3} \overline{m_i(\omega_1, \omega_2, \omega_3)}\, \tilde{m}_i(\omega_1, \omega_2 + \pi, \omega_3 + \pi) = 0 \end{cases}$$

The design of such filters is detailed in [CS]. Basically, one first choose the analyzing high pass $\tilde{m}_1$ and derive m_0 as a minor of the system (4.12) assuming that the global determinant is equal to one (to ensure that the solutions are FIR filters). This assumption leads to a duality relation that is used for the design of $\tilde{m}_0$ and the synthesis high pass m_1 is finally derived by solving the rest of the system.

A remarkable fact is that some particular choices of $\tilde{m}_1$ as $\tilde{m}_1(\omega_1, \omega_2, \omega_3) = \left(\frac{1 - e^{i\omega_1}}{2}\right)^2$ exactly leads to the low pass filter m_0 associated to the hat function $\varphi(x)$. The functions $\varphi, \tilde{\varphi}, \psi$ and $\tilde{\psi}$ are represented on the Figures 10 and 11. Note that $\tilde{\varphi}$ and $\tilde{\psi}$ are irregular but this is not so important since they are only used for the analysis and the reconstructing elements are splines.

Other examples of hexagonal bases are given in [CS]. It is shown in particular that this geometry provides a better angular resolution in the frequency plane than the tensor product. More precisely, we can divide the frequency band of a sampled signal on Γ, i.e. the hexagon

(4.13) $H = \{|\omega_1 - \omega_2| \le 2\pi\} \cap \{|\omega_2 - \omega_3| \le 2\pi\} \cap \{|\omega_3 - \omega_1| \le 2\pi\}$

into several channel representing a specific direction in frequency. The idea is to subdivide the high-pass subbands with the same filters. The result, shown on Figure 9 can be useful for edge detection and wavefront analysis in PDE's.

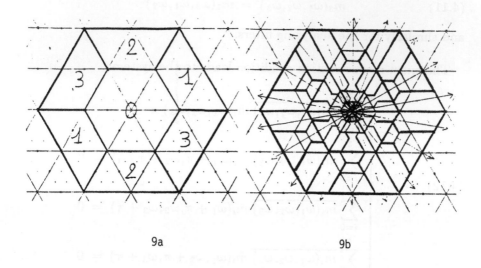

9a   9b

Figure 9

a) Hexagonal frequency decomposition
b) Example of subdivision for a sharper angular resolution

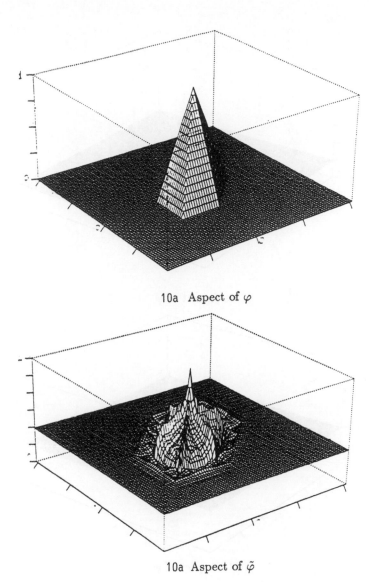

10a   Aspect of φ

10a   Aspect of $\tilde{\varphi}$

Figure 10

The scaling functions φ and $\tilde{\varphi}$ (Hexagonal)

24

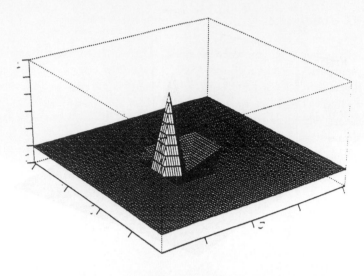

11a Aspect of ψ

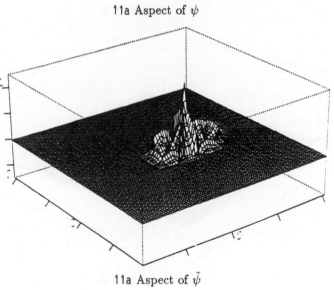

11a Aspect of $\tilde{\psi}$

Figure 11

The wavelets ψ and $\tilde{\psi}$ (Hexagonal)

V. Conclusion.

The new framework of biorthogonal wavelet bases seems very promising in the applied point of view : in the one dimensional case, its flexibility allows the use of almost any filter, provided that there exist a dual. This can be very useful for specific applications that requires filters which are not in the CQF (orthonormal) class. In the bidimensional case, it is a good framework for the design of non-separable filters which have a better isotropy than those obtained by the tensor product.

References.

[BA] P. Burt and E. Adelson, *"The Laplacian pyramid as a compact image code"*, IEEE Trans. Comm. vol. 31, pp. 482-540, 1983.

[CD1] A. Cohen and I. Daubechies, *"Non-separable bidimensional wavelet bases"*, submitted to Revista Matematica Iberoamericana, 1991.

[CD2] A. Cohen and I. Daubechies, *"A stability criterion for biorthogonal wavelets and their related subband coding schemes"*, preprint Bell Labs, 1991.

[CDF] A. Cohen, I. Daubechies and J.C. Feauveau, *"Biorthogonal bases of compactly supported wavelets"*, to appear in Comm. in Pure and Appl. Math., 1990.

[Co1] A. Cohen, *"Ondelettes, Analyses Multirésolutions et Filtres Miroirs en Quadrature, Ann. de l'IHP, Analyse Non Linéaire, vol. 7, pp. 439-459, 1990.

[Co2] A. Cohen, *"Biorthogonal Wavelets"*, in "Wavelets and Applications", C.K. Chui editor, Academic Press, 1991.

[CS] A. Cohen and J.M. Schlenker, *"Compactly supported wavelets with hexagonal symmetry"*, submitted to a special issue of Constructive Approximation on Wavelets, 1991.

[**Dau1**] I. Daubechies, *"Orthonormal Bases of Compactly supported Wavelets"*, Comm. Pure & Appl. Math., vol. 41, pp. 909-996, 1988.

[**Dau2**] I. Daubechies, *"Ten Lectures on Wavelets"*, notes of the CBMS Conference of Lowell (MA, USA), ed. Jones and Bartlett, 1991.

[**DL**] I. Daubechies and J. Lagarias, *"Two scale difference equations. Part I and II"*, to appear in SIAM Journ. Math. Anal., 1989.

[**JC**] J. Johnston and A. Cohen, *"Optimal dual filters for signal processing"*, preprint Bell Labs, 1991.

[**Ma**] S. Mallat, *"A theory for Multiresolution Signal Decomposition: The Wavelet Representation"*, IEEE PAMI 2 vol.7, 1989.

[**Me**] Y. Meyer, *"Ondelettes et Opérateurs"*, ed. Hermann, Paris, 1990.

[**SB**] M.J. Smith and T.P. Barnwell, *"Exact Reconstruction Techniques for Tree Structured Subband Coders"*, IEEE ASSP, vol. 34, pp. 434-441, 1986.

[**Ve**] M. Vetterli, *"Filter Banks allowing Perfect Reconstruction"*, Signal Processing, vol. 10, pp. 219-244, 1986.

Chapter 2

Nonrectangular Wavelet Representation of 2-D Signals. Application to Image Coding

Christine Guillemot*, A. Enis Cetin** and Rashid Ansari***

* CCETT, 4 rue du Clos Courtel, 35512 Cesson-Sévigné, FRANCE

** BILKENT University, Dept. of E.E. Eng., Bilkent, Ankara 06533, TURKEY

*** BELLCORE, 445 South Street, Morristown, NJ07962-1910, USA

Abstract

In image and video coding it is often efficient to use nonrectangular sampling grids and in particular the line quincunx grid for which the one-dimensional (1-D) wavelet representation is not applicable. In these applications, a two-dimensional (2-D) nonrectangular wavelet representation is desirable. Such a representation is implemented by using nonseparable filter banks in a tree structure. For example, in the case of quincunx sampling grids, filters with diamond-shaped passband responses are used. This chapter reviews the concepts of nonrectangular wavelet representation of two-dimensional signals and describes procedures for constructing families of orthogonal and biorthogonal 1-D and nonrectangular 2-D wavelet bases. The properties of phase linearity and regularity of these wavelet families are discussed, and their application to image coding is illustrated.

1. Introduction

The concept of Wavelet is introduced in 1984 by Goupillaud, Grossman and Morlet [1] as a new mathematical tool for multiresolution decomposition of continuous-time signals, and wavelets have since received wide attention both in the continuous and the discrete case [1-5]. This mathematical tool for multiresolution analysis of signals is investigated and applied in various fields including geophysics, image analysis for the purpose of segmentation, pattern recognition and coding. The incentive for this is its ability to provide a multiresolution or multiscale analysis of signals with flexible space-frequency localization. Wavelets are functions that are generated from one single function by translations and dilations, and compared to usual signal analysis techniques such as the Short-Time Fourier Transform (STFT), they have proved to be effective in performing both a fine spatial analysis and a fine frequency analysis, within the limits of the uncertainty principle, that provides the lower bound on the space-bandwidth product. They are thus convenient for the task of isolating discontinuities in a signal. The mathematical theory underlying wavelets is described in [5], and reviewed in a previous chapter of this book. In this chapter the emphasize is more on the concepts generalized to the two-dimensional non-separable case, and on practical solutions of wavelet families obtained relying on the filter bank theory.

The wavelet theory is closely related to the filter bank theory. As a matter of fact, it is shown in [6] that compactly supported orthonormal one-dimensional (1-D) wavelets with arbitrarily high regularity can be constructed using FIR filter design techniques. As pointed out in [6] these solutions are related to conjugate-quadrature filter solutions in a two-channel exact-reconstruction filter bank described by Smith and Barnwell [7]. Independently, a unifying framework, called multiresolution analysis, establishing the relationship between orthogonal pyramid-dyadic tree expansions of a signal [10, 11] and the wavelet theory [1-5], is defined [8,9]. The close connection between wavelets and multirate filter banks is also described in [12-14], and the notion of biorthogonal wavelets, by analogy with biorthogonal filter banks, is introduced [12, 15]. More recently, it is shown that the orthogonal wavelet solutions introduced in [6] are similar to filter bank solutions obtained by factorization of Lagrange halfband filters into conjugate-quadrature components [16,17], and to perfect reconstruction Quadrature Mirror Filters derived from Hermite polynomials [18]. In [16], it is also shown that the factorization of Lagrange halfband filters into linear-phase components with simple integer coefficients, provides a procedure for generating compactly supported biorthogonal wavelet bases.

The focus in the litterature is mainly on dyadic multiresolution representation of one-dimensional signals using one-dimensional wavelets. For the representation of multidimensional signals, the one-dimensional wavelet functions are applied separably by assuming rectangular sampling grids [8,9,18bis]. However in image representation it is often desirable to sample 2-D signals on nonrectangular sampling grids and in particular, on the line quincunx grid or on the hexagonal grid [19]. The multiresolution analysis of such signals requires the use of 2-D wavelet transforms, implemented with 2-D nonseparable filter banks, associated to nonrectangular sampling grids. In this chapter, the nonrectangular wavelet representation of 2-D signals is described and its application to image coding is illustrated. Instead of focusing on the mathematical formalism underlying the wavelet and multiresolution analysis theory, our interest here is primarily to describe the different concepts, and methods for deriving appropriate practical solutions of wavelet families. Considering 2-D continuous square integrable functions belonging to $L^2(\mathbf{R}^2)$, continuous families of nonseparable 2-D wavelets are first defined. We proceed to the discrete parameter case and define a discrete set, also called frame, of continuous wavelets for 2-D nonrectangular sampling grids using lattice theory [19]. The multiresolution analysis framework is then followed, leading to the definition of a sequence of embedded closed subspaces as successive approximations of functions of $L^2(\mathbf{R}^2)$ and to the definition of orthogonal subspaces containing the high frequency details or the components of the wavelet transform of functions of $L^2(\mathbf{R}^2)$. Based on the multiresolution analysis approach and on the multirate filter bank theory [24, 25], conditions for obtaining orthogonal and biorthogonal 2-D wavelet bases are derived.

A procedure for generating solutions of one-dimensional orthogonal and biorthogonal bases [16] is then described. The orthogonal bases are actually a special case of conjugate-quadrature filters (CQF) derived from Lagrange discrete-time interpolators, or halfband filters [26] using Lagrange interpolation coefficients. Biorthogonal wavelets can be constructed by factorizing Lagrange halfband filters into filters with linear phase frequency responses and simple coefficients, components of exact reconstruction filter banks without mirror symmetry. Orthogonal solutions have a nonlinear phase, whereas biorthogonal wavelets trade the orthogonality property for the phase linearity property.

The construction of wavelet bases for the representation of 2-D signals, combined with a quincunx-sampling grid is next addressed, and methods for deriving appropriate orthogonal and biorthogonal bases are described. This sampling grid is chosen due to the fact that in most naturally occuring scenes, the image information is largely confined to a diamond-shaped frequency band and information in this band is efficiently packed in the digital signal by choosing a quincunx grid. The first procedure based on generalized McClellan transformations [27] allows the derivation of solutions with features such as regularity, phase linearity, equiripple or maximally flat responses desirable for generating orthogonal and biorthogonal bases suited for the multiresolution representation of 2-D signals, on quincunx-sampling grids. The second method described, based on frequency transformations, suited for synthesizing IIR exact reconstruction filter banks [28], allows the construction of orthogonal bases with approximate linear-phase and infinite support, as well as compactly supported orthogonal bases. The nonrectangular wavelet representation finds interesting applications in image coding. To conclude this chapter results of coding using some of the solutions presented are discussed.

2. Nonrectangular Wavelet Representation of 2-D Signals

This section describes continuous and discrete families of functions called wavelets. The link between orthogonal and biorthogonal wavelet bases and pyramidal expansions of a signal using exact reconstruction filter banks is then reviewed.

2.1 Nonseparable Two-dimensional Wavelets

Let us define a 2-D wavelet family $\psi_{\mu,\underline{b}}(\underline{t})$ obtained from a 2-D function $\psi(\underline{t})$, called the mother wavelet, by translation and dilation/contraction in both dimensions as:

$$\psi_{\mu,\underline{b}}(\underline{t}) = \left| \det\mu \right|^{-1/2} \psi(\mu^{-1}\underline{t} - \mu^{-1}\underline{b}), \tag{1}$$

where $\underline{t} = \begin{bmatrix} t_1 & t_2 \end{bmatrix}^T$, $\underline{b} = \begin{bmatrix} b_1 & b_2 \end{bmatrix}^T \in \mathbf{R}^2$, and $\left| \det\mu \right|^{-1/2}$ is used for normalization. The signal scale conversion (or dilation/contraction) instead of being an isotropic factor as in the 1-D or separable case, is defined by the nonsingular matrix μ

$$\mu = \begin{bmatrix} \mu_{00} & \mu_{01} \\ \mu_{10} & \mu_{11} \end{bmatrix} \quad , \quad \mu_{ij} \in \mathbf{R}. \tag{2}$$

According to the values of μ_{ij}, the function $\psi_{\mu,\underline{b}}(\underline{t})$ is a stretched version or a contracted version of the mother wavelet $\psi(\underline{t})$. As a consequence of the scaling property of the Fourier transform, a stretched version of the mother wavelet in the spatial domain corresponds to a contracted version in the transform domain and inversely. Accordingly, these functions are short duration high frequency functions or long duration low frequency functions. This provides the flexibility in the space-frequency resolution of the analysis.

The continuous wavelet transform maps a function $f(\underline{t})$ onto a space-scale domain by expanding the function $f(\underline{t})$ in terms of the wavelet family. This expansion can be expressed by the inner product:

$$<\psi_{\mu,\underline{b}}(\underline{t}),f(\underline{t})> = \int_{\underline{t}\in R^2} \psi_{\mu,\underline{b}}(\underline{t}) \cdot f(\underline{t})d\underline{t} \tag{3}$$

The scaling and translation parameters being elements of R^2, this transform is highly redundant. To avoid this redundancy, the dilation or contraction parameter μ and the translation parameter $\underline{b}$ are discretized. The discretization must eliminate the redundancy while preserving the completeness property of the discrete set of wavelets in $L^2(R^2)$. The completeness property is essential in order to obtain orthonormal bases of $L^2(R^2)$. The discretization of these parameters is done by selecting a sequence of scales $(\mu_0^j)_{j\in Z}$ (corresponding to an exponential sampling of μ),

$$\mu=\mu_0^j, \tag{4}$$

and by uniformly sampling, for each scale, the translation vector $\underline{b}$ on a 2-D grid at a rate proportional to μ_0^j:

$$\underline{b}=\mu_0^j.\mathbf{B}.\underline{n}, \tag{5}$$

where $j \in \mathbf{Z}$, represents the index of the scale conversion and $\underline{n}\in Z^2$. The scaling matrix, μ_0 is a nonsingular 2x2 matrix with integer coefficients and $\mathbf{B}$ is also an integer 2x2 matrix. In the following we will assume $\mathbf{B} = \mathbf{I}$, where $\mathbf{I}$ is the identity matrix. Note that μ^{-j} is the product of $|j|$ matrices μ^{-1} where $\mu \cdot \mu^{-1} = \mathbf{I}$. The matrix μ_0 represents the elementary scaling unit or dilation step and the global scaling matrix varies discretely by integer powers of μ_0. A discrete set of continuous wavelet functions $\psi_{j,\underline{n}}(\underline{t})$ is obtained, and expressed as:

$$\psi_{j,\underline{n}}(\underline{t}) = \left| \det\mu_0 \right|^{-j/2} \psi(\mu_0^{-j}\underline{t}-\underline{n}) = S_j T_{\underline{n}}\psi. \tag{6}$$

where S_j stands for scale conversion by a power j of the matrix μ_0^{-1} along with normalization and $T_{\underline{n}}$ denotes the translation by the integer vector $\underline{n}$. If this set is complete in $L^2(R^2)$ for given ψ, μ_0 and $\underline{b}$, then the continuous wavelet functions $\psi_{j,\underline{n}}(\underline{t})$ are called affine wavelets or frames. More details about the mathematical properties of affine wavelets or frames can be found in [51,52]. A frame is *tight* if

$$\sum_m \sum_{\underline{n}} |<f,\psi_{m,\underline{n}}>|^2 = ||f||^2. \tag{7}$$

To be linearly independant and constitute a basis in $L^2(R^2)$, the frame must also be *exact* or *minimal*, meaning that removing one of the functions of the frame results in a non complete set of wavelet functions. A tight exact frame constitutes an orthonormal basis and satisfy:

$$<\psi_{j\underline{n}_1},\psi_{j\underline{n}_2}> = \delta_{\underline{n}_1\underline{n}_2} \tag{8}$$

where $\delta_{\underline{n}_1\underline{n}_2}$ is the Kronecker symbol and is equal to "1" if the two vectors are identical and to "0" otherwise. This condition of orthogonality between the functions on a same scale is obviously met if the basis functions are of finite extent and their translates do not overlap. Moreover, the scaled versions of ψ must also be orthogonal across the scales and satisfy

$$<\psi_{j_1\underline{n}},\psi_{j_2\underline{n}}> = \delta_{j_1j_2}. \tag{9}$$

The decomposition of a function $f(\underline{t})$ belonging to $L^2(R^2)$ in the orthonormal basis $\psi_{j,\underline{n}}$ defines the nonrectangular discrete 2-D wavelet transform of the function $f(\underline{t})$, and is noted as:

$$W_f(j,\underline{n}) = <f , S_j T_{\underline{n}}\psi> = \left| \det\mu_0 \right|^{-j/2} \int_{\underline{t}\in R^2} f(\underline{t}) \cdot \psi(\mu_0^{-j} \cdot \underline{t} - \underline{n}) \cdot d\underline{t}. \tag{10}$$

2.2 Nonrectangular Multiresolution Representation

The concept of multiresolution analysis [8,9] provides a general approach in the construction of wavelets. It allows to establish the link between orthogonal / biorthogonal wavelets and pyramidal expansions of a signal using exact reconstruction filter banks. This section reviews this approach by extending it to nonrectangularly sampled two-dimensional signals.

2.2.1 Multiresolution Analysis and Orthogonal Wavelets

Multiresolution analysis has been defined as a framework in which square integrable functions $f \in L^2(R^d)$ are considered as a limit of successive approximations. Each approximation is a smoothed version of the original function f at different resolutions, obtained by convolution of the function $f(\underline{t})$ by a lowpass function $\phi(\underline{t})$ called the smoothing function. The difference between two successive approximations is the high frequency detail function.

Let us consider continuous square integrable 2-D functions belonging to $L^2(R^2)$. Let the nonsingular matrix μ_0 with determinant M be the scaling parameter between two successive approximations. Given μ_0, a successive nonrectangular approximation of a function $f \in L^2(R^2)$ is defined by the sequence of embedded closed subspaces V_j, $j \in Z$ satisfying:

a) $V_{j+1} \subset V_j \quad j \in Z,$ (11)

b) $\lim_{j \to -\infty} V_j$ is dense in $L^2(R^2)$ and $\lim_{j \to \infty} V_j = \left\{ 0 \right\},$ (12)

c) For all $j \in Z$ if $f \in V_j$ then $S_1 f \in V_{j+1}$ or if $f(\underline{x}) \in V_j$ then $f(\mu_0^{-1} \underline{x}) \in V_{j+1}$.   (13)

d) There exists a scaling function $\phi(\underline{t}) \in V_0$ such that:

for all $j \in Z$, the set $\dfrac{1}{\sqrt{M^j}} \phi(\mu_0^{-j} \underline{t} - \underline{n}), \underline{n} \in Z^2$ is an orthonormal basis of V_j.   (14)

Let W_{j+1} be the subspace orthogonal to V_{j+1} and complementary of V_{j+1} in V_j. The sequence of subspaces W_{j+1} represents the difference between two successive approximations V_j and V_{j+1}. This sequence satisfies the conditions:

a) W_{j+1} is orthogonal $V_{j+1} \quad j \in Z,$ (15)

and

b) $W_{j+1} \bigcirc V_{j+1} = V_j.$ (16)

The direct sum of all the spaces W_j spans $L^2(R^2)$. There exists a function ψ, which by scaling by μ_0^{-j} and translating on a 2-D grid defined by LAT(μ^{-j}), generates the set of functions:

$$\frac{1}{\sqrt{M^j}} \psi(\mu_0^{-j} \underline{t} - \underline{n}) , \underline{n} \in \mathbf{Z}^2, \tag{17}$$

that constitutes an orthonormal basis of W_j. Since the subspaces W_j are mutually orthogonal and the direct sum of all the spaces W_j spans $L^2(\mathbf{R}^2)$, it follows that

$$\left\{ \frac{1}{\sqrt{M^j}} \psi(\mu_0^{-j} \underline{t} - \underline{n}) \right\}_{j \in \mathbf{Z}, \underline{n} \in \mathbf{Z}^2} \quad \text{is an orthonormal basis of } L^2(\mathbf{R}^2).$$

Let A_j be the orthogonal projection operator from $L^2(\mathbf{R}^2)$ onto V_j. The approximation of f at the resolution V_j is given by the orthogonal projection of f onto V_j, which can be written as a linear combination of the basis functions $\phi(\mu_0^{-j}.\underline{t} - \underline{n})$:

$$A_j f(\underline{t}) = \frac{1}{\sqrt{M^j}} \sum_{\underline{n}} < f, S_j T_{\underline{n}} \phi > . \phi(\mu_0^{-j}.\underline{t} - \underline{n}) \tag{18}$$

$$= \sum_{\underline{n}} a_{j\underline{n}}(f).\phi(\mu_0^{-j}.\underline{t} - \underline{n}).$$

$A_j f(\underline{t})$ represents an approximation of f at the resolution M^{-j}, and provides the smoothness or low frequency information content of the function $f(\underline{t})$. Since $\lim_{j \to \infty} V_j$ is dense in $L^2(\mathbf{R}^2)$, it follows that $\lim_{j \to \infty} A_j f(\underline{t}) = f(\underline{t})$, for all $f \in L^2(\mathbf{R}^2)$.

Let D_j be the orthogonal projection operator from $L^2(\mathbf{R}^2)$ onto W_j. The projection D_j of a function $f \in L^2(\mathbf{R}^2)$ onto W_j is defined as the linear combination of the basis functions $\psi(\mu_0^{-j}.\underline{t} - \underline{n})$:

$$D_j f(\underline{t}) = \frac{1}{\sqrt{M^j}} \sum_{\underline{n}} < f, S_j T_{\underline{n}} \psi > . \psi(\mu_0^{-j}.\underline{t} - \underline{n}) \tag{19}$$

$$= \sum_{\underline{n}} d_{j\underline{n}}(f) \psi(\mu_0^{-j}.\underline{t} - \underline{n}).$$

$D_j f(\underline{t})$ represents the detail signal or high frequency information content of $f(\underline{t})$. Since $V_j = W_{j+1} + V_{j+1}$, it follows that $A_j f = A_{j+1} f + D_{j+1} f$. As a result, since $V_0 = \text{Span}[\phi(\underline{t}-\underline{n})]$ and $V_{-1} = \text{Span}[\phi(\mu_0^{-1}\underline{t}-\underline{n})]$ it follows that $W_0 = \text{Span}[\psi(\underline{t}-\underline{n})]$. The functions $\psi(\mu_0^{-j}.\underline{t} - \underline{n})$ are wavelet functions as defined in the previous section. In summary, the set of inner products:

$$\left\{ a_{j\underline{n}}(f), d_{j\underline{n}}(f), \underline{n} \in \mathbf{Z}^2 \right\} \tag{20}$$

represents the *nonrectangular wavelet representation* of a 2-D function belonging to $L^2(\mathbf{R}^2)$, and consists of the signal at a coarser resolution $a_{j\underline{n}}(f)$ and of the detail signals

$d_{j\underline{n}}(f)$ at successive resolutions μ^{-j}.

Let us now focus on finding appropriate smoothing and wavelet funstions satisfyimg all the properties described above. Since $V_1 \subset V_0$ and $W_1+V_1 = V_0$, smoothing and wavelet functions $\phi(\underline{t})$ and $\psi(\underline{t})$ at resolution j=0 can be expressed in terms of the smoothing function at resolution j=1 by:

$$\phi(\underline{t}) = \sum_{\underline{n} \in Z^2} g_0(\underline{n})\phi(\mu_0^{-1}\underline{t}-\underline{n}) \quad \text{and} \quad \psi(\underline{t}) = \sum_{\underline{n} \in Z^2} g_1(\underline{n})\phi(\mu_0^{-1}\underline{t}-\underline{n}). \tag{21}$$

The coefficients $g_0(\underline{n})$ and $g_1(\underline{n})$ can be identified as the coefficients of respectively a lowpass kernel and a highpass kernel, and are called interscale coefficients. These relations between resolutions j and j+1 can be written as:

$$<\phi_{j+1\underline{n}},\phi_{j\underline{n}}> = g_0(\underline{n}) \quad \text{and} \quad <\psi_{j+1\underline{n}},\phi_{j\underline{n}}> = g_1(\underline{n}). \tag{22}$$

Taking the Fourier transforms of these expressions gives:

$$\Phi(\underline{\Omega}) = G_0(e^{j\mu_0\underline{\omega}}) \cdot \Phi(\mu_0\underline{\Omega}) \quad \text{and} \quad \Psi(\underline{\Omega}) = G_1(e^{j\mu_0\underline{\omega}}) \cdot \Phi(\mu_0\underline{\Omega}) \tag{23}$$

with $\underline{\omega} = \underline{\Omega} \cdot \underline{T} = [\omega_1 \ \omega_2]^T$, being the discretized normalized frequency. By iterating for all the resolutions, assuming $\Phi(\underline{0}) = 1$, it can be shown that:

$$\Phi(\underline{\Omega}) = \prod_{k=1}^{\infty} G_0(e^{j\mu_0^k\underline{\omega}}) \quad \text{and} \quad \Psi(\underline{\Omega}) = G_1(e^{j\mu_0^k\underline{\omega}})\prod_{k=1}^{\infty} G_0(e^{j\mu_0^k\underline{\omega}}) \tag{24}$$

Note that if $g_0(\underline{n})$ and $g_1(\underline{n})$ are of finite duration, then $\phi(\underline{t})$ and $\psi(\underline{t})$ are compactly supported.

The problem to be solved now is to find the conditions to be satisfied by $g_0(\underline{n})$ and $g_1(\underline{n})$ so that the sets $\phi_{j\underline{n}}(\underline{t})$ and $\psi_{j\underline{n}}(\underline{t})$ constitute orthonormal bases of respectively V_j and W_j. Since $\phi(\underline{t}-\underline{n})$ and $\psi(\underline{t}-\underline{n})$ are orthonormal bases spanning V_0 and W_0 their Fourier transforms satisfy the unitary conditions:

$$\sum_{\underline{\Omega}_i \in P(\mu_0)} |\Phi(\underline{\Omega}+\underline{\Omega}_i)|^2 = 1 \quad \text{and} \quad \sum_{\underline{\Omega}_i \in P(\mu_0)} |\Psi(\underline{\Omega}+\underline{\Omega}_i)|^2 = 1 \tag{25}$$

leading to:

$$\sum_{\underline{\omega}_i \in P(\mu_0)} |G_0(\underline{\omega}+\underline{\omega}_i)|^2 = 1 \quad \text{and} \quad \sum_{\underline{\omega}_i \in P(\mu_0)} |G_1(\underline{\omega}+\underline{\omega}_i)|^2 = 1 \tag{26}$$

where $P(\mu_0) = \left\{ \underline{n} \in Z^2 : \mu_0^{-1}\underline{n} \in [0,1)^2 \right\}$ [29]. Since the subspaces V_j and W_j are orthogonal, the Fourier transforms of the functions $\phi(\underline{t})$ and $\psi(\underline{t})$ satisfy:

$$\left| \Phi(\underline{\Omega}) \right|^2 = \sum_{j=1}^{J} \left| \Psi(\mu_0^j \underline{\Omega}) \right|^2 + \left| \Phi(\mu_0^J \underline{\Omega}) \right|^2, \tag{27}$$

Substituting (23) and (25) in (27) gives:

$$|G_0(e^{j\underline{\omega}})|^2 + |G_1(e^{j(\underline{\omega})})|^2 = 1 \tag{28}$$

The orthogonality condition between scaling and wavelet functions,

$$<\phi_{\underline{mn}}(\underline{t}), \psi_{\underline{kl}}(\underline{t})> = 0 \tag{29}$$

can be rewritten in the Fourier transform domain as:

$$\sum_{\underline{\Omega}_i \in P(\mu_0)} \phi(\mu_0\underline{\Omega} + \underline{\Omega}_i) \cdot \overline{\psi(\mu_0\underline{\Omega} + \underline{\Omega}_i)} = 0, \tag{30}$$

where $\overline{\psi}$ denotes the complex conjugate of ψ. Substituting (23) and (25) in (30) gives:

$$\sum_{\underline{\omega}_i \in P(\mu_0)} G_0(\underline{\omega} + \underline{\omega}_i)\overline{G_1(\underline{\omega} + \underline{\omega}_i)} = 0. \tag{31}$$

In most naturally occuring scenes, the image information is largely confined to a diamond-shaped frequency band and information in this band is efficiently packed in the digital signal by choosing a quincunx grid. Thus, in the following sections the emphasis is on the particular case where the scaling matrix μ_0 is given by:

$$\mu_0 = \begin{bmatrix} 1 & 1 \\ 1 & -1 \end{bmatrix}, \tag{32}$$

and $M = |\det\mu_0| = 2$. Conditions (26), (28) and (31) can be rewritten as:

$$|G_0(\underline{\omega})|^2 + |G_0(\underline{\omega}+\underline{\Pi})|^2 = 1, \tag{33}$$

$$|G_0(\underline{\omega})|^2 + |G_1(\underline{\omega})|^2 = 1, \tag{34}$$

and,

$$G_0(\underline{\omega})\overline{G_1(\underline{\omega})} + G_0(\underline{\omega} + \underline{\Pi})\overline{G_1(\underline{\omega} + \underline{\Pi})} = 0, \tag{35}$$

where $\underline{\Pi}=[\pi\ \pi]^T$. These conditions are equivalent to the requirement that the matrix

$$\begin{bmatrix} G_0(\underline{\omega}) & G_1(\underline{\omega}) \\ G_0(\underline{\omega} + \underline{\Pi}) & G_1(\underline{\omega}+\underline{\Pi}) \end{bmatrix}, \tag{36}$$

be a unitary matrix. The conditions imposed on the filters g_0 and g_1 so that $\phi(\underline{t})$ and $\psi(\underline{t})$ are respectively smoothing and wavelet functions, generating orthonormal basis in $L^2(\mathbf{R}^2)$, appropriate for the multiresolution analysis of 2-D signals are equivalent to the exact reconstruction conditions of a two-channel self-transpose filter bank as shown in Figure 1, where g_0 and g_1 are required to be *power-complementary* filters (34). On the synthesis side, the synthesized signal is approximated as a linear combination of dilated and shifted version of ψ. The discrete wavelet transform is thus recognized to be equivalent to a tree filter bank. The novelty of the wavelet theory comes down to the choice of the filters which must satisfy the *regularity property*. This notion is explained in section 2.2.3.

Figure 1 : Two-Channel Nonseparable Analysis/Synthesis Filter Bank

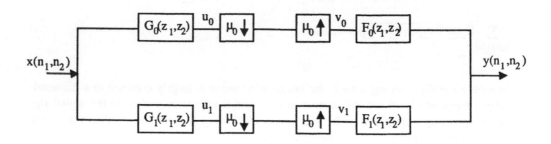

2.2.2 Multiresolution Analysis and Biorthogonal Wavelets

The condition of orthogonality is very restrictive in the choice of filters satisfying the exact reconstruction property. In particular, there is no FIR solution with linear phase. For applications such as image coding, where the features of finite impulse response and phase linearity are desired, the constraint of orthogonality between the lowpass and highpass kernels g_0 and g_1 of the analysis bank is relaxed, and a *biorthogonal* analysis/synthesis filter bank pair is defined. To satisfy the exact reconstruction condition an orthogonality constraint is imposed across the analysis and synthesis sections of the analysis/synthesis pair. To illustrate this notion, consider the two-channel filter bank pair shown in Figure 1. The signals u_k and the upsampled signals v_k, $k = 0,1$, are related by

$$v_k[\underline{n}] = \begin{cases} 2u_k[\underline{n}] & n_1 + n_2 = \text{even} \\ \\ 0 & n_1 + n_2 = \text{odd} \end{cases}$$

$$= (1 + (-1)^{n_1 + n_2})\, u_k[\underline{n}]. \tag{37}$$

The corresponding z-transforms are related by

$$V_k(\underline{z}) = [U_k(\underline{z}) + U_k(-\underline{z})]. \tag{38}$$

Looking at Figure 1 and using (38), the z-Transform of the output y can be expressed as

$$Y(\underline{z}) = [F_0(\underline{z})\, G_0(\underline{z}) + F_1(\underline{z})\, G_1(\underline{z})]\, X(\underline{z})$$

$$+ [F_0(\underline{z})\, G_0(-\underline{z}) + F_1(\underline{z})\, G_1(-\underline{z})]\, X(-\underline{z}). \tag{39}$$

The first term on the right hand side is proportional to $X(\underline{z})$ and represents the amplitude distorsion, and the second term containing the aliasing component $X(-\underline{z})$ arises due to a frequency translation caused by downsampling and upsampling. This undesirable aliasing term in the reconstructed signal vanishes if the filters are chosen to satisfy the condition given by

$$F_0(\underline{z})\, G_0(-\underline{z}) + F_1(\underline{z})\, G_1(-\underline{z}) = 0. \tag{40}$$

The reconstruction is exact and the filters are *biorthogonal* if the first term satisfies:

$$F_0(\underline{z}) \, G_0(\underline{z}) + F_1(\underline{z}) \, G_1(\underline{z}) = 1. \tag{41}$$

In the multiresolution analysis framework this is equivalent to defining two sequences of embedded closed subspaces:

$$V_{j+1} \subset V_j \quad j \in Z, \quad \text{and} \quad V_{j+1} \subset \overline{V}_j, \, j \in Z \tag{42}$$

as well as two sequences of orthogonal complement subspaces W_j and $\overline{W}_j$ such that W_j is complementary to V_j in V_{j-1} and orthogonal to $\overline{V}_j$, and $\overline{W}_j$ is complementary to $\overline{V}_j$ in $\overline{V}_{j-1}$ and orthogonal to V_j. Two families of wavelet coefficients along with corresponding interscale coefficients can be defined:

$$\phi(\underline{t}) = \sum_{\underline{n} \in Z^2} g_0(\underline{n}) \phi(\mu_0^{-1}\underline{t}-\underline{n}) \tag{43}$$

$$\overline{\phi}(\underline{t}) = \sum_{\underline{n} \in Z^2} \overline{f}_0(\underline{n}) \overline{\phi}(\mu_0^{-1}\underline{t}-\underline{n}) \tag{44}$$

$$\psi(\underline{t}) = \sum_{\underline{n} \in Z^2} g_1(\underline{n}) \psi(\mu_0^{-1}\underline{t}-\underline{n}) \tag{45}$$

$$\overline{\psi}(\underline{t}) = \sum_{\underline{n} \in Z^2} \overline{f}_1(\underline{n}) \overline{\psi}(\mu_0^{-1}\underline{t}-\underline{n}) \tag{46}$$

for which the following relations of orthogonality hold:

$$<\overline{\phi}_{m\underline{n}} , \psi_{m'\underline{n}'}> = 0 , \quad \text{and} \quad <\overline{\psi}_{m\underline{n}} , \phi_{m'\underline{n}'}> = 0 \quad m,m' \in Z , \quad \underline{n},\underline{n}' \in Z^2, \tag{47}$$

$$<\overline{\psi}_{m\underline{n}} , \overline{\psi}_{m'\underline{n}'}> = \delta_{mm'}\delta_{\underline{n}\underline{n}'} , \quad <\overline{\phi}_{m\underline{n}} , \overline{\phi}_{m\underline{n}'}> = \delta_{\underline{n}\underline{n}'} \text{ and } <\overline{\psi}_{m\underline{n}} , \overline{\psi}_{m\underline{n}'}> = \delta_{\underline{n}\underline{n}'} \tag{48}$$

Procedures for designing linear phase filter banks and biorthogonal wavelets satisfying the above relations and that are suitable for 1-D signal representation and 2-D nonrectangular representation are described in section 3.

2.2.3 Regularity or Smoothness Property

In order to represent the signal at many resolutions, the convolution of the smoothing function with the analyzed signal is iterated for a number of stages j. One important condition on the filter g_0 is that for $j \rightarrow \infty$, the function $g_0^{(j)}(\underline{n})$, ($g_0^{(j)}(\underline{n})$ denoting the iterated convolution of the impulse response g_0 by itself), converges toward a continuous function and not toward a discontinuous or fractal function. In other words,

$$\phi(\omega) = \lim_{k \to \infty} \prod_k G_0(\mu_0^{-k}\omega) \tag{49}$$

where $\phi(\omega)$ is a continuous function. This condition ensuring that no "artificial" discontinuity (not due to the analyzed signal) ever appears in the analysis is called the *regularity property*. The scaling function $\phi(t)$ and the associated lowpass kernel g_0 are regular. From (21) it follows that all regularity properties derived for $\phi(t)$ carry over $\psi(t)$, thus the study of regularity can be restricted to the function $\phi(t)$. This condition also guarantees that the discrete wavelet transform characterized by the smoothing and wavelet filters g_0 and g_1 converges toward the continuous wavelet transform.

Mathematical analysis in the one-dimensional case of the implications of this regularity feature also called smoothness property on the filter responses are provided in [6], [13], [14], [51] and [53], and is summarized here. The regularity order is given by the number of times the function is continuously differentiable, or $d^N\phi(t)/dt^N$ is continuous. This is a smoothness requirement on the spatial waveforms of $g_0^{(j)}(n)$. This requirement has direct implications on the characterization of the filters g_0 and g_1. The continuity of $d^N\phi(t)/dt^N$ implies the uniform convergence of $\delta^N g_0^{(j)}(n)$. In order to have this expression uniformly convergent, it is shown that $G_0(z)$ must have at least $N+1$ zeroes at $z=-1$ [6], [52]. Imposing N zeroes at $z=-1$ for $G_0(z)$ amounts to requiring that the frequency response $|G_0(e^{j\omega})|$ is flat about half the sampling frequency $\omega = \pi$. This condition is necessary but not sufficient, since other zeroes in $G_0(z)$ may have a destructive effect of the zeroes located at $z=-1$. The amount of regularity lost due to the effect of other zeroes is characterized by the *Holder regularity* [53] stipulating that g_0^j is regular with order $r=N+\alpha$ if $|\delta^N g_{0_n}^j - \delta^N g_{0_n}^j| < c2^{-j\alpha}$. The maximum negative value of α for which this expression holds is the amount of regularity lost due to the destructive effect of other zeroes. The overall regularity order is then $N - \alpha$.

In the two-dimensional case, it is conjectured that zeroes at the points of replication of the spectrum have a favorable effect for regularity. The design methods described in section 3, when choosing regular one-dimensional prototypes, generate solutions with a sufficient number of zeroes at the points of replication of the spectrum. It is verified that these solutions are highly regular.

3. Construction of Bases for the Representation of Quincunx-Sampled Signals

This section describes methods for constructing practical solutions of wavelet functions adapted to the multiresolution representation of 2-D signals, on quincunx sampling grids. These procedures are based on transformations applied to one-dimensional prototypes.

3.1 1-D Wavelet Construction using Lagrange Halfband Filters

Before addressing the problem of constructing nonrectangular 2-D bases for the wavelet representation of quincunx signals, the construction of 1-D wavelet bases that will serve as the building blocks for the quincunx signal representation is examined. It is shown that compactly supported orthonormal bases described in [6] can be derived from Lagrange interpolation filters [16, 17]. These wavelet filters can be used to derive separable wavelet bases for representation of images, and can also be used, as shown in the following sections, as prototype filters in the synthesis of nonrectangular 2-D wavelet filters.

3.1.1 Lagrange Halfband Filters

Here the use of Lagrange halfband filters in the construction of a two-channel exact-reconstruction filter bank with both linear-phase and CQF responses is described. The results obtained for the CQF case turn out to be identical to those leading to the wavelet solutions in [6]. The transfer function H(z) of a halfband symmetric FIR filter [26] is given by

$$H(z) = \tfrac{1}{2} + z\, T(z^2). \tag{50}$$

Note that even samples of the impulse response are constrained such that $h[0] = \tfrac{1}{2}$ and $h[2n] = 0$ for $n = \pm1, \pm2, \cdots$. If the filter has an even-symmetric impulse response, with $h[n] = h[-n]$, then the transfer function of a halfband symmetric FIR filter is given by

$$H_K(z) = \tfrac{1}{2} + \sum_{n=1}^{K} h_K[2n-1](z^{-2n+1} + z^{2n-1}). \tag{51}$$

The subscript K is used to indicate that the filter has duration 4K−1. The frequency response is purely real and can be made non-negative by suitably scaling the odd samples of the impulse response. Halfband filters are useful in decimation and interpolation by factors of 2. A special case of halfband filters is obtained by choosing the filter coefficients according to the Lagrange interpolation formula [30], and the filter coefficients are given by:

$$h_K[2n-1] = \frac{(-1)^{n+K-1} \prod_{i=1}^{2K}(K + \tfrac{1}{2} - i)}{(K-n)!\,(K-1+n)!\,(2n-1)}, \tag{52}$$

where $n = 1, 2, ..., K$. The frequency response in this case is non-negative. Such a halfband filter is useful in deriving CQF solutions for filter banks that can be used in the construction of wavelets described in [6]. The impulse response for the case of K = 4 is shown in

Table 1.

Table 1 : Halfband Filter Impulse Response Values.
Impulse Response $h_4[n]$, $h_4[n] = h_4[-n]$, (with a scaling factor / 4096)

n	0	1	2	3	4	5	6	7
$h_4[n]$	2048	1225	0	-245	0	49	0	-5

It can be shown that these Lagrange halfband filters are a special case of maximally-flat digital filters [31-34]. An efficient program for obtaining coefficients in the more general case of maximally-flat digital filters with varying balance between the degrees of flatness at zero frequency and the flatness at the Nyquist (half-sampling) frequency is described in [33]. For the special case of the maximally-flat halfband filter a closed form expression for the ratio of filter coefficients $h_K[2n-1]/h_K[1]$ is given in [34]. With some manipulations it can be verified that those filter coefficients are identical to the Lagrange halfband filter coefficients in (52). A factorization of Lagrange halfband filter to construct exact-reconstruction filter banks is now considered.

3.1.2 Solutions for Orthogonal and Biorthogonal Bases

Consider the two-channel analysis/synthesis filter bank pair shown in Figure 2. The z–transform of the output y can be expressed as

$$Y(z) = [F_0(z) G_0(z) + F_1(z) G_1(z)] X(z)$$

$$+ [F_0(z) G_0(-z) + F_1(z) G_1(-z)] X(-z). \tag{53}$$

Figure 2 : Two-Channel Analysis/Synthesis Filter Bank

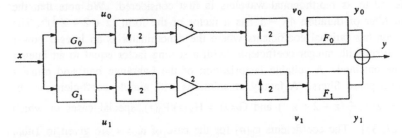

Filters satisfying conditions for exact reconstruction at the output, i.e. $Y(z) = z^{-n_0}X(z)$, can be obtained by choosing a factorization of the halfband filter transfer function $H_K(z)$ as

$$H_K(z) = F_0(z) \, G_0(z). \tag{54}$$

The frequency response $H_4(e^{j\omega})$ is shown in Figure 3, with the filter impulse response given in Table 1. A proper combination of factors can be chosen to get the filters $G_0(z)$ and $F_0(z)$. The synthesis filters $G_1(z)$ and $F_1(z)$ can be suitably defined so that conditions for exact-reconstruction are satisfied.

Figure 3 : Magnitude Response of Filters $H_4(e^{j\omega})$, Linear-Phase (LP) Filter $G_0(e^{j\omega})$ and CQF Filter $G_0(e^{j\omega})$ for K = 4.

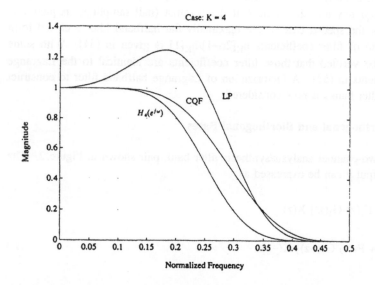

A partition of Lagrange filters into linear-phase factors which lead to wavelet solutions that are referred to as biorthogonal wavelets is first considered. We note that the Lagrange halfband filter of duration $4K - 1$ has a factor of the form $(z + 2 + z^{-1})^K$. The filter transfer function has rational coefficients where the denominator is an integer power of 2. As a result filters with integer coefficients (with a scaling factor equal to an integer power of 2) can be obtained. A suitable factorization of the Lagrange halfband transfer function yields linear-phase filters. A family of solutions with simple coefficients is obtained by letting $F_0(z) = \frac{1}{4}(z + 2 + z^{-1})$ and $G_0(z) = H_K(z)/F_0(z)$, special cases of which are described in [27, 35]. The coefficients $g_0[n]$ for the case of K = 4 are given in Table 2. The linear-phase frequency response $G_0(e^{j\omega})$ is plotted with the label LP in Figure 3 for K = 4.

Table 2 : Linear-Phase Filter Impulse Response Values (LP).
$g_0[n] = g_0[-n]$ (Scaling Factor / 1024).

n	0	1	2	3	4	5	6
$g_0[n]$	700	324	-123	-78	34	10	-5

The construction of a conjugate-quadrature filter bank using Lagrange halfband filters is now considered. This construction gives rise to orthogonal wavelets. As mentioned before, the filter frequency response is non-negative allowing the transfer function to be factored as

$$H_K(z) = G_0(z)\, G_0(z^{-1}).$$
(55)

The CQF solution $G_0(z)$ for the case K = 4 is given in Table 3 in its factored form with terms that include (i) a factor $(z + 2 + z^{-1})^2$, (ii) a second-order factor with zeros inside the unit circle, and (iii) a first-order factor also with its zero inside the unit circle. The frequency response is shown in Figure 3 with the label CQF. The filter coefficients of $G_0(z)$ for the case K = 4 are given in Table 4. Note that $G_0(1) = 1$ in first column of Table 4. The second column shows the filter coefficients scaled up by a factor of $\sqrt{2}$ which can be seen to be the same as those obtained by Daubechies [6] (see column 3) in obtaining wavelet solutions with regularity conditions. The CQF solutions for K = 1,2,...,10 were obtained by using MATLAB and the solutions are identical to those appearing in [6]. It has been shown [36] that the filters $H_K(z)$ can also be derived by taking a linear combination of filters called binomial filters and solving for the coefficients by imposing the perfect reconstruction condition. Here we have a closed form solution for the halfband filter coefficients $h_K[n]$ according to (52).

Table 3 : Factors of CQF Transfer Function $G_0(z)$, for K = 4.
Note: $H_k(z) = G_0(z)G_0(z^{-1})$.

$(0.25z + 0.50 + 0.25z^{-1})^2$
$(1.74923621306475.z - 0.99390306558956 + 0.24466685252481.z^{-1})$
$(1.49003742601680 - 0.49003742601680.z^{-1})$

44

Table 4 : Coefficients of CQF Transfer Function for K = 4.

$G_0(z)$	$\sqrt{2}\,G_0(z)$	Daubechies' result
0.16290171402565	0.23037781330890	0.230377813309
0.50547285754591	0.71484657055291	0.714846570553
0.44610006912338	0.63088076792986	0.630880767930
−0.01978751311782	−0.02798376941686	−0.027983769417
−0.13225358368452	−0.18703481171909	−0.187034811719
0.02180815023709	0.03084138183556	0.030841381836
0.02325180053549	0.03288301166689	0.032883011667
−0.00749349466518	−0.01059740178507	−0.010597401785

Regular linear-phase one-dimensional wavelets can also be derived, by setting the perfect reconstruction condition as a Bezout identity [13]. The solutions are then obtained using Diophantine equations and continued fractions. Linear-phase filter banks with some control over the choice of the filter can also be constructed. For example, by constraining the lowpass filter to be a halfband filter, the following exact reconstruction filter bank is defined [37]:

$$G_0(z) = H(z) , \tag{56}$$

$$F_0(z) = 2z^{-1} + [1 - \frac{1}{2}zT(z^2)]\frac{1}{2}T(z^2) , \tag{57}$$

$$G_1(z) = \frac{1}{2}F_0(-z) , \tag{58}$$

and

$$F_1(z) = 2G_0(-z), \tag{59}$$

This choice of filters allow to generate a family of biorthogonal wavelets, with some control on the smoothing function.

3.2 Construction of Nonrectangular 2-D Orthogonal Bases

Consider the analysis/synthesis filter bank pair as shown in Figure 1. Equations (40) and (41) give the conditions on the filters so that the analysis/synthesis pair constitutes an exact reconstruction system. In addition, the orthogonality condition requires that the filters g_0 and g_1 are power-complementary filters. Thus, the following relations hold between their z-transforms:

$$\left| G_0(\underline{z}) \right|^2 + \left| G_1(\underline{z}) \right|^2 = 1 \ , \quad \underline{z} = \left[e^{j\omega_1} \ \ e^{j\omega_2} \right]^T, \tag{60}$$

and

$$G_0(\underline{z}) \cdot G_1(\underline{z}^{-1}) + G_0(-\underline{z}) \cdot G_1(-\underline{z}^{-1}) = 0, \tag{61}$$

with $\underline{z} = [z_1 \ z_2]^T$ and $\underline{z}^{-1} = [z_1^{-1} \ z_2^{-1}]^T$. One possible solution is given when requiring that the highpass filter be obtained by a frequency translation of the lowpass filter (with a possible delay) so as to get a pair of mirror image responses:

$$g_1(\underline{n}) = (-1)^{n_1 + n_2} g_0(\underline{n} - \underline{N}), \tag{62}$$

for some $\underline{N} \in Z^2$. The wavelet filter is the modulated version by $(-1)^{n_1 + n_2}$ of the smoothing filter, and the following relation holds between their z-transforms

$$G_1(z_1, z_2) = z_1^{-K_1} z_2^{-K_2} G_0(-z_1^{-1}, -z_2^{-1}), \tag{63}$$

where K_1, K_2 are integers. Under the restriction (63), (40) can be satisfied by choosing $K_1 + K_2 = $ odd, together with

$$F_0(z_1, z_2) = G_0(z_1^{-1}, z_2^{-1}) \tag{64}$$

and

$$F_1(z_1, z_2) = z_1^{K_1} z_2^{K_2} G_0(-z_1, -z_2). \tag{65}$$

The overall system with input x and output y now becomes a time invariant system. The z-Transforms $Y(\underline{z})$ and $X(\underline{z})$ are related by

$$Y(\underline{z}) = C(\underline{z}) \ X(\underline{z}). \tag{66}$$

In (66) $C(\underline{z})$ is given by

$$C(\underline{z}) = H(\underline{z})H(\underline{z}^{-1}) + H(-\underline{z})H(-\underline{z}^{-1}), \tag{67}$$

where $\underline{z}^{-1}$ for convenience is used to denote $[z_1^{-1}, z_2^{-1}]^T$, and

$$H(z_1, z_2) = G_0(z_1, z_2). \tag{68}$$

3.2.1 Compactly Supported Bases

Linear Phase Solutions:

It is first assumed that $H(\underline{z})$ is a linear-phase filter. To have exact reconstruction, it is required that $Y(\underline{z}) = X(\underline{z})$ or $Y(\underline{z}) = z_1^{-K_1} z_2^{-K_2} X(\underline{z})$, the output being a delayed version of the input signal. Let us require that $Y(\underline{z}) = X(\underline{z})$ for simplicity of notation, then we need $C(\underline{z}) = 1$. At this point it is assumed that $H(\underline{z})$ is a linear phase FIR filter with a real impulse that satisfies the following centro-symmetry conditions:

$$h[n_1, n_2] = h[m-n_1, m-n_2], \quad m = 0, 1, \tag{69}$$

and

$$h[n_1, n_2] = h[1-m-n_1, m-n_2] \quad m = 0, 1, \tag{70}$$

that correspond to respectively same and different symmetry, odd or even, in the two dimensions. Noting that any transfer function $H(\underline{z})$ can be expressed in the polyphase form as follows

$$H(z_1, z_2) = H_0(z_1 z_2, z_1^{-1} z_2) + z_2^{-1} H_1(z_1 z_2, z_1^{-1} z_2), \tag{71}$$

the condition (67) can be rewritten as

$$\tfrac{1}{2}[H_0(z_1 z_2, z_1^{-1} z_2)H_0(z_1^{-1} z_2^{-1}, z_1 z_2^{-1}) + H_1(z_1 z_2, z_1^{-1} z_2)H_1(z_1^{-1} z_2^{-1}, z_1 z_2^{-1})] = 1. \tag{72}$$

This equation represents the power-complementary property satisfied by the polyphase components.

Claim: The centro-symmetry conditions admit only trivial solutions of no more than two filter taps.

Proof:

Taking the z-transform of both sides of (69) gives:

$$H(z_1,z_2) = z_1^{-m}z_2^{-m}H(z_1^{-1},z_2^{-1}) \tag{73}$$

implying that

$$H_1(z_1z_2,z_1^{-1}z_2) = H_0(z_1^{-1}z_2^{-1},z_1z_2^{-1}). \tag{74}$$

Similarly, considering m=0 or m=1 in (70), it can easily be shown that the polyphase components under condition (70) are also related by (74).

As a result, (67) can be rewritten as:

$$H_0(z_1z_2,z_1^{-1}z_2)H_0(z_1^{-1}z_2^{-1},z_1z_2^{-1}) = 1. \tag{75}$$

Since $H_0(\underline{z})$ is the transfer function of an FIR filter, (75) forces H_0 to have only one non-zero sample. So, under the constraint of linear-phase finite impulse response, (67) has only simple two taps solutions given by:

$$h(\underline{n}) = \begin{cases} 1/2 & \underline{n}=[L_1 \ L_2]^T,[1-L_1,-L_2]^T \\ 0 & \text{otherwise} \end{cases} \tag{76}$$

where L_1,L_2 are integers. It appears clearly that the choice of filters as in (76) is not suitable for a diamond-shaped response.

Non Linear-phase Solutions:

If the constraint of phase linearity is relaxed, solutions to (60) and (61) are given by even length Conjugate Quadrature Filters (CQF):

$$G_1(\underline{z}) = z_1^{k_1}z_2^{k_2}G_0(-\underline{z}^{-1}) \ , \ (k_1,k_2 \in \mathbf{Z}, k_1 + k_2 \ \text{odd}). \tag{77}$$

A procedure for designing Fan filters [28] is used here to synthesize a filter bank using diamond-shaped CQFs. The method is based on a rotational frequency transformation of one-dimensional prototype filters. The prototype filters considered are CQF one-dimensional filters [7], and one-dimensional wavelets as described in section 3.1. The transfer function of a 2-D rectangular-shaped low-pass filter can be derived from a one-dimensional low-pass prototype filter h by the equation

$$D_{lp}(z_1, z_2) = H(z_1)H(z_2) + H(-z_1)H(-z_2), \qquad (78)$$

where $H(z)$ is the z-transform of the prototype filter h. Let $T_1^{(h)}(z)$ and $T_2^{(h)}(z)$ be the polyphase components of the one-dimensional prototype filter h. The one-dimensional transfer function $H(z)$ can be expressed in the form

$$H(z) = T_1^{(h)}(z^2) + zT_2^{(h)}(z^2). \qquad (79)$$

A rotational frequency transformation given by the change of variables,

$$z_1 \leftarrow z_1^{1/2} \cdot z_2^{1/2} \qquad (80)$$

and

$$z_2 \leftarrow z_1^{1/2} \cdot z_2^{-1/2} \qquad (81)$$

applied to $D_{lp}(z_1, z_2)$ allows to obtain the transfer function $G_0(\underline{z})$ of the 2-D diamond-shaped filter, to be used as the lowpass kernel in the exact reconstruction analysis/synthesis pair. This transfer function can be expressed as

$$G_0(z_1, z_2) = T_1^{(h)}(z_1 z_2) \cdot T_1^{(h)}(z_1 z_2^{-1}) + z_1 \cdot T_2^{(h)}(z_1 z_2) \cdot T_2^{(h)}(z_1 z_2^{-1}). \qquad (82)$$

Similarly, the application of the rotational frequency transformation to the separable high-pass filter having for transfer function

$$D_{hp}(z_1, z_2) = G(-z_1)G(z_2) + G(z_1)G(-z_2), \qquad (83)$$

where $G(z)$ is the z-transform of the one-dimensional filter g conjugate quadrature of h, generates the diamond-shaped high-pass filter $G_1(\underline{z})$, conjugate quadrature of $G_0(\underline{z})$. The transfer function of $G_1(\underline{z})$ is given by

$$G_1(z_1, z_2) = T_1^{(g)}(z_1 z_2) \cdot T_1^{(g)}(z_1 z_2^{-1}) - z_1 \cdot T_2^{(g)}(z_1 z_2) \cdot T_2^{(g)}(z_1 z_2^{-1}), \tag{84}$$

where $T_1^{(g)}$ and $T_2^{(g)}$ are the polyphase components of the one-dimensional filter g. As the one-dimensional prototype filters h and g are conjugate quadrature, the following relations hold between their respective polyphase components:

$$T_1^{(g)}(z) = -z^{-2k-1} T_2^{(h)}(z^{-1}), \tag{85}$$

and

$$T_2^{(g)}(z) = z^{-2k'-1} T_1^{(h)}(z^{-1}).$$

It follows that the z-transforms of the 2-D diamond-shaped filters verify

$$G_1(\underline{z}) = -z_1^{-2K-1} G_0(-\underline{z}^{-1}). \tag{86}$$

By choosing the synthesis low-pass and high-pass filters such as

$$F_0(\underline{z}) = G_0(\underline{z}^{-1}) \tag{87}$$

and

$$F_1(\underline{z}) = z_1^{K_1} z_2^{K_2} G_0(-\underline{z}), \tag{88}$$

it is easy to verify that the conditions for aliasing cancellation (61) and perfect reconstruction (60) are satisfied.

Let us consider an example of design of such a filter bank. The prototype filter chosen is the one-dimensional filter with 8 taps given in [7]. The resulting 2-D response of the filter g_0 is plotted in Figure 4. Orthogonal bases can also be derived applying the rotational frequency transformation to maximally flat filters (which are consequently very regular) issued from the factorization of Lagrange interpolation filters as described in section 3.1. Note that the procedure can also be used to derive linear-phase diamond-shaped FIR filters from the 1-D even length filters proposed in [38] and from the 1-D odd length filters proposed in [39]. But these filters allow only to approximate the orthogonality condition (67) and they, consequently, cannot be used to generate orthogonal wavelet bases.

Figure 4 : $\left|G_0(\omega)\right|$ for Orthogonal Wavelet Bases, obtained by transformation from the 8-taps filter given in [7].

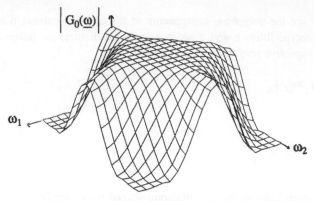

Let us examine the regularity of solutions derived by such a procedure. From the nature of the decomposition, to generate a wavelet basis, it is necessary to have regular low-pass filters. The wavelet transform is, as a matter of fact, obtained by iterating the branch of the decomposition having the low-pass filter as shown in (49). The iterated filtering and sampling process converges toward a continuous function if the filter is regular, otherwise it converges toward a fractal function.

Figure 5. Impulse Response of the smoothing function after the sixth iteration (a) using the filter derived by transformation of the [1 3 3 1] 1-D filter and (b) using the filter derived by transformation of the [1 -3 -3 1] 1-D filter.

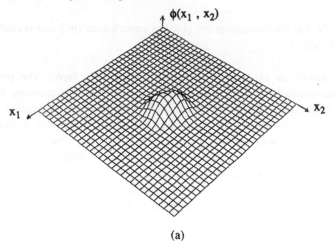

(a)

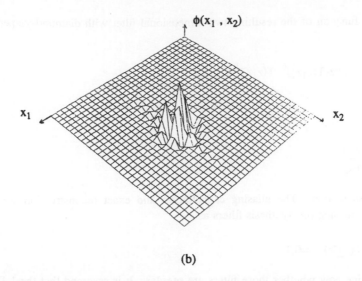

$\phi(x_1, x_2)$

x_1 x_2

(b)

As an example, the one-dimensional filters [1 3 3 1] and [1 -3 -3 1] given in [13] as respectively regular and non regular filters, by the transformation described in (78-88) result into respectively regular and non regular 2-D diamond-shaped low-pass and high-pass filters. Figures 5 (a) and (b) show the impulse responses obtained after the sixth iteration of filtering and subsampling using those two diamond-shaped filters. The cascading of the filtering and subsampling operations using the filter derived from the transformation of the [1 3 3 1] filter converges toward a continuous function, whereas the other one converges toward a fractal function.

3.2.2 Bases derived from IIR Filters

To obtain orthogonal diamond-shaped wavelet filters with approximately linear phase, a class of IIR filters [28], [40], which are mirror-image IIR filters are considered. These filters are obtained by applying the transformation given by (78-88) to a one-dimensional prototype filter with polyphase components

$$T_1(z) = 1, \tag{90}$$

and

$$T_2(z) = \prod_{k=1}^{L} \frac{a_k z + 1}{z + a_k}. \tag{91}$$

52

The transfer function of the resulting two-dimensional filter with diamond-shaped responses is given by

$$G_i(\underline{z})=1/2[1+(-1)^i z_1 T(z_1 z_2) \cdot T(z_1 z_2^{-1})] \quad , \quad i=0,1, \tag{92}$$

where,

$$T(z) = \prod_{k=1}^{L} \frac{a_k z+1}{z+a_k} \tag{93}$$

is an all-pass section. The aliasing cancellation and exact reconstruction conditions are satisfied by choosing the synthesis filters as

$$F_i(\underline{z}) = z_1 z_2 G_i(-\underline{z}^{-1}) \quad , \quad i=0,1. \tag{94}$$

Let us examine now whether those filters are regular. It is assumed that the 1-D prototype filter is regular of order N. Its z-transform has at least N zeroes at $z = -1$. The frequency transformation used in the synthesis procedure "displaces" these zerpes to different locations in the 2-D plane, including the points of replication of the spectrum. By choosing a prototype one-dimensional filter with a sufficient regularity order, the procedure guarantees a regular two-dimensional filter.

Figure 6: (a) Transfer Function of the low-pass filter g_0, (b) Transfer Function of the smoothing Function after the 6th iteration.

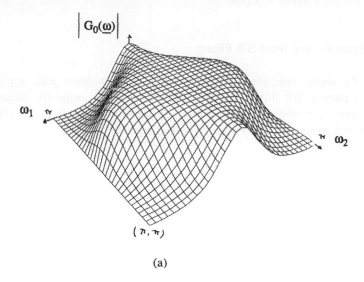

(a)

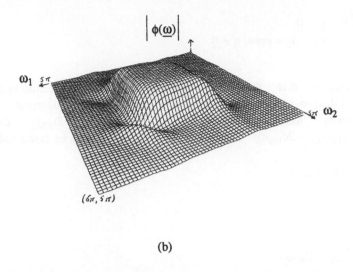

$$\left|\phi(\underline{\omega})\right|$$

ω_1 ˢπ

ˢπ ω_2

$(6\pi, 5\pi)$

(b)

Figure 6 shows the transfer function obtained from the sixth iteration of filtering with the diamond-shaped lowpass filter $g_0(n_1,n_2)$ with parameters { L=1 and $a_1=1/3$ } followed by subsampling. As it can be observed from this figure, the product $\prod_{k=1}^{6}G_0(\mu_0^{-k}\underline{\omega})$ is continuous.

3.3 Construction of Biorthogonal Bases

The biorthogonality of a nonseparable diamond-shaped system, as defined in [12] for the 1-D systems, is characterized by the following equation for the nonrectangular 2-D signals:

$$G_0(\underline{z})F_0(\underline{z}) + G_1(\underline{z})F_1(\underline{z}) = 1, \tag{94}$$

where $\underline{z} = \begin{bmatrix} z_1 & z_2 \end{bmatrix}^T$.

Solutions to this equation are provided by the filters obtained from the factorization of a half band diamond-shaped filter, imposing moreover a symmetry between the analysis and synthesis filters [27]. As a matter of fact, let $D_{HB}(\underline{z})$ be the transfer function and $d_{HB}(\underline{n})$ be the impulse response of a diamond halfband filter. The impulse response $d_{HB}(\underline{n})$ satisfies the following condition:

54

$$d_{HB}(\underline{n}) = \begin{cases} \frac{1}{2} & \underline{n} = \underline{0} \\ 0 & n_1 + n_2 = \text{even}, \ \underline{n} \neq \underline{0}. \end{cases} \tag{95}$$

It is assumed that the filter response $d_{HB}(\underline{n})$ is centro-symmetric with $d_{HB}(n_1,n_2) = d_{HB}(-n_1,n_2) = d_{HB}(-n_1,-n_2)$. We impose this requirement since diamond filters ideally have this symmetry. Note that $D_{HB}(\underline{z}) = D_{HB}(\underline{z}^{-1})$ and $D_{HB}(\underline{z}) + D_{HB}(-\underline{z}) = 2d_{HB}(\underline{0})$. The transfer function $D_{HB}(\underline{z})$ can be factorized as

$$D_{HB}(\underline{z}) = F_0(\underline{z}) \ G_0(\underline{z}). \tag{96}$$

Now let

$$G_1(\underline{z}) = z_1^{-K_1} z_2^{-K_2} F_0(-\underline{z}), \tag{97}$$

where $K_1 + K_2 = $ odd and

$$F_1(\underline{z}) = z_1^{K_1} z_2^{K_2} G_0(-\underline{z}). \tag{98}$$

Then it can easily be verified that:

$$F_0(\underline{z}) \ G_0(\underline{z}) + F_1(\underline{z}) \ G_1(\underline{z}) = D_{HB}(\underline{z}) + D_{HB}(-\underline{z})$$
$$= 2d_{HB}(\underline{0}) = 1. \tag{99}$$

and

$$F_0(\underline{z})G_0(-\underline{z}) + F_1(\underline{z})G_1(-\underline{z}) = 0 \tag{100}$$

Both the biorthogonality and the aliasing cancellation conditions are thus verified.

The problem is now how to get two-dimensional diamond-shaped halfband filters and factorize them. Factorizing two-dimensional polynoms is a very involved problem. Here, the key to the solution consists in using a 1-D to 2-D transformation that is applied to a *1-D halfband filter* of order $4N + 3$ with transfer function $H_{HB}^1(z)$ expressed as

$$H_{HB}^1(z) = h_{HB}^1(0) + \sum_{k=0}^{N} h_{HB}^1(2k+1)\,(z^{2k+1} + z^{-2k-1}). \tag{101}$$

Noting that

$$z^{2k+1} + z^{-2k-1} = 2T_k(\tfrac{1}{2}(z + z^{-1})), \tag{102}$$

where T_k is the Chebyshev polynomial of degree k, $H_{HB}^1(z)$ can be expressed as

$$H_{HB}^1(z) = h_{HB}^1(0) + \sum_{k=0}^{N} a_k(z+z^{-1})^{2k+1}. \tag{103}$$

In order to get a halfband filter with a diamond-shaped response the following transformation is used:

$$z + z^{-1} \;\leftarrow\; \tfrac{1}{2}(z_1 + z_1^{-1} + z_2 + z_2^{-1}). \tag{104}$$

The above transformation is McClellan-type [41] with frequency mapping given in Figure 7.

Figure 7: Contour Plot of 1-D to 2-D frequency mapping (parametrized by $\cos\omega/2 = 1, 0.5, 0, -0.5, -1$).

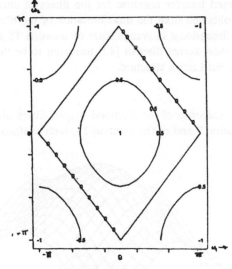

In summary, to derive the diamond-shaped components of the factorization of the halfband diamond-shaped filter, we consider the factorization components of a 1-D half band filter, to which we apply the transformation given by (104), and the frequency mapping represented in Figure 7. As an example, we designed a pair of filters generating biorthogonal bases by factorization of the 1-D Lagrange filter of duration 15 into filters of duration 3 and 13 and transformation of the respective filters. The coefficients of the one-dimensional component filters are given in table 5.

Table 5: Filter Response Values (scaling factor 4096), h[0], h[1], ...; h[n] = h[-n], of the 1-D Halfband Lagrange Filter of duration 15.

h(0)	h(1)	h(2)	h(3)	h(4)	h(5)	h(6)	h(7)
2048	1225	0	-245	0	49	0	-5

Table 6: Filter Response Values of the components of duration 3 and 13 (scaling factors 4, 1024) after factorization of the 1-D Halfband Lagrange Filter of duration 15.

(a)	2	1					
(b)	700	324	-123	-78	34	10	-5

The transformed transfer response for the filters of duration 3 and 13 is plotted in Figure 8. The filter obtained from the transformation of the filter of duration 15 derived by factorizing the one-dimensional Lagrange filter of duration 15 as explained previously, and also described as a short kernel filter in [42] turns out to be the same solution as obtained in [23] using a cascaded Lattice structure.

Figure 8. Frequency response of the diamond-shaped filters obtained by transformation of the filters of (a) duration 3 and of (b) duration 13, with coefficients given in Table 6.

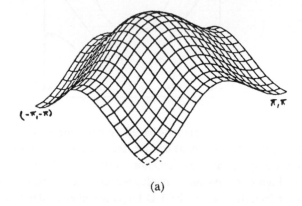

(a)

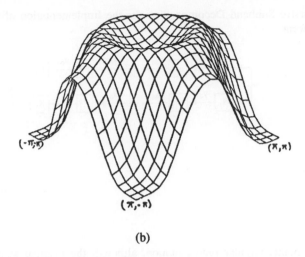

$(-\pi,\pi)$ (π,π)

$(\pi,-\pi)$

(b)

The application of the frequency transformation to a function having a zero at $\omega=\pi$ results in a 2-D function with zeros in $(0,0)$, $(0,\pi)$, $(\pi,0)$ and (π,π). By choosing a one-dimensional prototype filter with a sufficient number of zeros at $\omega=\pi$ the resulting 2-D function is ensured to have zeros at the points of replication of the 2-D spectrum, providing some degree of regularity. The transformation used in the design procedure has, thus, the nice property of conserving features of the 1-D prototype filters, such as phase linearity and regularity.

4. Application to Image Coding

In this section, the application of the nonrectangular wavelet representation to image coding is described. The wavelet representation can be implemented as a recursive subband decomposition as shown in Figure 9. This structure creates a hierarchy of sub-images of different frequency content and thus of different perceptual importance [43]. Coding schemes, based on such a structure are also known as layered coding. They allow to address desirable features for service integration, which are compatibility and inter-working between equipments. Various researchers [43-47] have considered layered coding schemes for video transmission on ATM networks, as the coding can be tailored to each sub-image which can also be assigned appropriate transmission priority. In the context of transmission on ATM networks, the layered structure naturally offers a wide range of qualities and bit rates, along with amenability to solutions for error concealment that guarantee some degree of robustness in case of packet loss. In terms of compression efficiency, the wavelet representation, as the subband decomposition, provides a smooth spatial overlap between the basis functions and thus do not suffer the drawback of producing blocking artifacts in the picture for low bit rate transmission.

Figure 9. Recursive Subband Decomposition for the Implementation of the Nonrectangular
Wavelet Transform

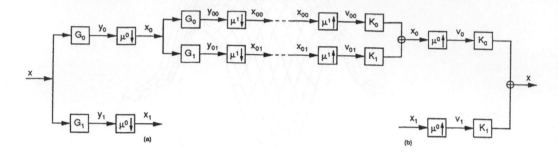

For the nonrectangular representation, although the original signal is rectangularly
sampled, the sub-signals at the subsequent resolutions are quincunx-sampled, the following
ones rectangularly sampled and so forth. Figure 10 shows the decomposition in the frequency domain. The results described here concern still image coding, but we will describe
later how they can be used for video. Those results have been obtained by using the
wavelets derived from IIR filters, with $T(z)$ of length $L=1$, referring to the notations used
in section 3.2.2. In the examples described here, two filter banks corresponding to
coefficient values $a_1 = 1/3$ and $a_1 = 1/4$ have been used. These two filters are suited for
fast implementation since the filter coefficients are integers. The IIR filters have approximately linear-phase characteristics in their passbands so that the lower resolution signals do
not exhibit phase distortion associated with IIR filters.

Figure 10. Decomposition in the Frequency Domain.

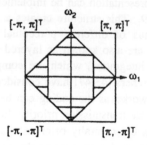

The lowest frequency subsignal, at each step of the decomposition contains most of
the signal energy and significant information, and also represents a lower resolution version
of the original image. This subimage contains the most significant perceptual information
and it is important to code it efficiently. This signal has been coded using a Discrete
Cosine Transform based technique [48]. The transform coefficients are scanned in a zig-zag

manner, and are quantized by rounding of suitably scaled coefficients. A large portion of the quantized coefficients are observed to be zero. The values of the non-zero coefficients are coded with an amplitude lookup table, and their locations are coded using a runlength table.

The high frequency subsignals are coded using first a deadzone quantizer for data compression. These subsignals are then coded noiselessly by employing runlength coding for runs of zero values using a lookup table and the non-zero values are amplitude coded using a Huffman code based lookup table. The quantizer can be chosen according to the statistics of the subsignals and their visual impact. It was found that a coarser quantizer can be used in the outermost band subsignal and a finer quantizer can be used in the inner high frequency band subsignal. The deadzones can also be chosen in a similar manner, i.e., the deadzone of the quantizer which codes the outermost band subsignal is larger than the quantizer of the inner high frequency band subsignal. However for more demanding images such as HDTV signals, one may have to use smaller deadzones. The quantization levels used beyond the deadzone were almost uniform with slightly smaller quantization intervals used for low magnitude samples. After the deadzone quantization, most of the test images have a significantly reduced number of non-zero values. The information on the location of these non-zero values can be efficiently coded using runlength coding [40].

The coding results for the test images known as Barbara and Kiel Harbour of respective sizes, 672x576, and 720x576, are reported in Tables [7-8]. Among the parameters changed were the filter coefficients (a_1 = 1/3 filter #1 and 1/4 filter #2), the quantization of outer bands, the DCT block size, and the rounding procedure. These results are comparable to those obtained in the case of rectangular subband coding [39]. The main advantage of nonrectangular processing is the hierarchy created by the partition into two subbands without any horizontal or vertical bias and filter characteristics matching those of human visual perception. This feature is attractive in situations where there is a loss of the high band information due to reasons such as network congestion, in the context of video transmission over ATM networks. It may also allow a definition of quantization laws more adapted to the human visual perception. To see the effect of the loss of the high band information, the image Barbara was applied to the analysis filter bank and the outermost band was discarded. The remaining information was used to reconstruct the signal. The results appeared visually more acceptable than in the case where either the high vertical or high horizontal information in the rectangular subband decomposition was ignored. The SNR for this image is 30.3 dB whereas the SNR for this image when the high horizontal rectangular subband is discarded is of only 27.2 dB. The orthogonal property of wavelets is an interesting feature in the case of deliberate loss of the high frequency band, for low bit rate coding, or accidental loss, in the case of packet loss in an ATM environment. Even when the high frequency bands are discarded, the aliasing cancellation is preserved.

Compact supported wavelets described in section 3.2.1 can be applied similarly for image coding. They have the orthogonality property, but their phase is not linear. The biorthogonal wavelets described in section 3.3, can also be used for image coding, following the same tree structure decomposition.

So far, we have discussed the use of the nonrectangular wavelet representation for still image coding. Nevertheless, the representation can also be used for video coding, for conventional Television (TV) or High Definition TV (HDTV) applications. As a matter of fact, the structure, by nature, addresses the issue of compatibility. The term compatibility here implies the resolution compatibility, also called downward and upward compatibility, and also that a decoder conforming to a specific syntax, should be able to decode an easily extractable stream from the information intended for a higher quality decoder, a feature that is sometimes referred as backward compatibility [49]. Forward compatibility refers to the feature that allows a higher resolution decoder to accept and decode a lower resolution bit stream. This structure can thus be very useful in the definition of compatible TV/HDTV coding schemes. Moreover, the core encoder/decoder, used to code/decode the lower resolution signal, can be compatible to existing standards, as for example the MPEG I syntax as described in [50], where a rectangular subband decomposition is used, or the future MPEG II standard.

Summary

In this chapter, the concepts of nonrectangular wavelet representation have been described as well as procedures for designing practical solutions of orthogonal and biorthogonal bases. These methods in the case of one-dimensional wavelet functions are based on the factorization of Lagrange halfband filters. The methods providing two-dimensional solutions for the representation of signals on quincunx-sampled grids are based on frequency transformations. Solutions with different characteristics of phase linearity, finite or infinite support, are described. Their behaviour in terms of regularity is also examined. The application of nonrectangular wavelet representation of 2-D signals to image coding is also illustrated.

References

1 P. Goupillaud, A. Grossman and J. Morlet, "Cycle-Octave and Related Transforms in Seismic Signal Analysis," *Geoexploration*, Vol. 23, pp. 85-102, Elsevier Science Publishers, B.V. Amsterdam, Netherlands, 1984/1985.

2 A. Grossmann and J. Morlet, "Decomposition of Hardy functions into Square Integrable Wavelets of Constant Shape," *SIAM J. Math. Anal. 15 (1984)* pp. 723-736.

3 I. Daubechies, A. Grossman and Y. Meyer, "Painless nonorthogonal expansions," *J. Math. Phys.*, 27, pp. 1271-1283, 1986.

4 S. Jaffard, "Construction of Wavelets on Open Sets," *Proceedings Conf. Wavelets, Time-Frequency Methods and Phase Space*, Lecture Notes on IPTI, Springer-Verlag, New-York, 1989.

5 Y. Meyer, "Ondelettes et Opérateurs," Vol. 1, Ed. Hermann, Paris, 1990.

6 I. Daubechies, "Orthogonal Bases of Compactly Supported Wavelets", *Communication*

on Pure and Applied Mathematics, vol. XLI, pp. 909–996, 1988.

7 M. J. T. Smith and T. P. Barnwell, "Exact Reconstruction Techniques for Tree-Structured Subband Coders", *IEEE Trans. on Acoust., Speech, Signal Processing*, vol. ASSP–34, no. 3, pp. 434–441, June 1986.

8 S. G. Mallat, "A Theory for Multiresolution Signal Decomposition: The Wavelet Representation, " *IEEE Trans. on Pattern Analysis and Machine Intelligence*, Vol II, no 7, pp. 674-693, July 1989.

9 S. G. Mallat, "Multifrequency Channel Decompositions of Images and Wavelet Models," *IEEE Trans. on Acoust., Speech, Signal Proc.*, Vol. 37, No. 12, Dec. 1989.

10 E. H. Adelson, E. Simoncelli, R. Hingorani, "Orthogonal Pyramid Transforms for Image Coding", *Proc. SPIE Vol 845 Visual Communications and Image Processing II*, pp. 50-58, 1987.

11 J. W. Woods, "Subband Image Coding," *Ed. Kluwer Academic Publishers*, 1990.

12 M. Vetterli and C. Herley, "Wavelets and Filter Banks: Relationships and New Results," *Proc. Int. Conf. Acoust., Speech and Signal Proc.*, Albuquerque, Vol. NM, pp. 1723-1726, 1990.

13 M. Vetterli and C. Herley, "Wavelets and Filter Banks: Theory and Design," *Submitted to IEEE Trans. Acoust., Speech, Signal Proc.*, To appear Sept. 1992.

14 O. Rioul, " A Unifying Multiresolution Theory for the Discrete Wavelet Transform, Regular Filter Banks and Pyramid Transforms," *submitted to IEEE trans. on Acoust., Speech, Signal Proc.*, June 1990,

15 A. Cohen, I. Daubechies and J.C Feauveau, "Biorthogonal Bases of Compactly Supported Wavelets," *Technical memo. #11217-900529-07*, AT&T Bell Laboratories, Murray Hill, NJ, USA.

16 R. Ansari, C. Guillemot and J. Kaiser, "Wavelet Construction Using Lagrange Interpolation Filters," *IEEE Trans. Circuits and Systems*, pp. 1116-1118, Sept. 1991.

17 M. J. Shensa, "The Discrete Wavelet Transform: Wedding the à Trous and Mallat Algorithms," *IEEE Trans. Acoust., Speech, Signal Proc.*, To appear, Oct. 1992.

18 A. N. Akansu, R. A. Haddad, and H. Caglar, "Perfect Reconstruction Binomial QMF-Wavelet Transform," *Proc. SPIE Visual Communication and Image Processing*, vol. 1360, pp. 609-618, Oct. 1990.

19 E. Dubois, "The Sampling and Reconstruction of Time-Varying Imagery with Application in Video Systems," *Proc. IEEE*, vol. 73, pp. 502-522, April 1985.

20 J. C. Feauveau, "Analyse Multirésolution pour les Images avec un Facteur de Resolution $\sqrt{2}$," *Revue de traitement du signal, Vol. 7, no. 2*, pp. 117-128, 1990.

21 A. E. Cetin, "A Multiresolution Nonrectangular Wavelet Representation For Two-Dimensional Signals," *Proc. 1990 Bilkent Int. Conf.*, pp. 1432-1438, July 1990.

22 C. Guillemot, A. E. Cetin and R. Ansari, "M-Channel Nonrectangular Wavelet Representation for 2-D Signals," *Proc. IEEE Intern. Conf. on Acoust., Speech, and Signal Proc.*, pp. 2813-2816, Toronto, May 1991.

23 J. Kovacevic and M. Vetterli, " Nonseparable Multidimensional Perfect Reconstruction Filter Banks and Wavelet Bases for $\mathbf{R^n}$," *IEEE Trans. Acoust., Speech, Signal Proc.*, Submitted for Publication, February 1991.

24 M. Vetterli, "Multirate Filter Banks for Subband Coding," Chapter in *Subband Image*

Coding, Ed. J. W. Woods, Kluwer Academic Publishers, 1990.

25 P. P. Vaidyanathan, "Quadrature Mirror Filter Banks, M-Band Extensions and Perfect Reconstruction Techniques," *IEEE ASSP Magazine,* vol. 4, no. 3, pp. 4-20, July 1987.

26 M. G. Bellanger, J. L. Daguet, and G. P. Lepagnol, "Interpolation, Extrapolation and Reduction of Computation Speed in Digital Filters", *IEEE Trans. Acoust., Speech, Signal Processing,* vol. ASSP-22, no. 4, pp. 231-235, August 1974.

27 R. Ansari and C. Guillemot, "Exact Reconstruction Filter Banks Using Diamond FIR Filters", *Proc. 1990 Bilkent Int. Conf.,* Elsevier Press, Amsterdam, The Netherlands, pp. 1412–1424, July 1990.

28 R. Ansari and C. L. Lau, "Two-dimensional IIR Filters for Exact Reconstruction in Tree-structured Subband Decomposition", *Electronics Letters,* Vol 23, June 1987.

29 R. Ansari and S. Lee, "Two-Dimensional Nonrectangular Interpolation, Decimation and Filter Banks", *Presented at Int. Conf. on Acoust., Speech, Signal Proc.,* 1988.

30 M. Abramowitz and I. A. Stegun, Editors, *Handbook of Mathematical Functions,* National Bureau of Standards, U.S. Government Printing Office, Washington, D.C. 20402, 1970.

31 O. Herrmann, "On the Approximation Problem in Nonrecursive Digital Filter Design", *IEEE Trans. Circuit Theory,* vol. CT-18, no. 2, pp. 411-413, May 1971; reprinted in *Digital Signal Processing,* edited by L. R. Rabiner and C. M. Rader, IEEE Press, New York, NY, pp. 202-203, 1972.

32 J. F. Kaiser and W. A. Reed, "Data Smoothing Using Low Pass Digital Filters", *Review of Scientific Instruments,* vol. 48, no. 11, pp. 1447-1457, November 1977.

33 J. F. Kaiser, "Design Subroutine (MXFLAT) for Symmetric FIR Low Pass Digital Filters with Maximally-Flat Pass and Stop Bands", Section 5.3, *Programs for Digital Signal Processing,* IEEE Press, New York, pp. 5.3-1 to 5.3-6, 1979.

34 C. Gumacos, "Weighting Coefficients for Certain Maximally Flat Nonrecursive Digital Filters", *IEEE Trans. Circuits and Systems,* vol. CAS-25, no. 4, pp. 234-235, April 1978.

35 R. Ansari and D. J. Le Gall, "Advanced Television Coding Using Exact Reconstruction Filter Banks", Chapter 7, *Subband Image Coding,* J. W. Woods, Ed., Kluwer Academic Publishers, Norwell, MA, pp. 297-298, 1991.

36 A. N. Akansu, R. A. Haddad and H. Caglar, "Perfect Reconstruction Binomial QMF-Wavelet Transform", *Proc. SPIE Conf. Visual Commun. and Image Processing,* Lausanne, Switzerland, vol. 1360, pp. 609–618, Octobre 1990.

37 C. W. Kim, R. Ansari and A. E. Cetin, "Biorthogonal Linear-Phase Halfband Wavelets," to appear in *Proc. IEEE int. Conf. Acoust., Speech, Signal Proc.,* 1992.

38 J. D. Johnston, "A Filter Family Designed for Use in Quadrature Mirror Filter Banks", *Proc. IEEE int. Conf. Acoust., Speech, Signal Proc.,* pp.291-293, 1980.

39 E. H. Adelson, E. Simoncelli, R. Hingorani, "Orthogonal Pyramid Transforms for Image Coding", *Proc. SPIE Vol 845 Visual Communications and Image Processing II,* pp. 50-58, 1987.

40 R. Ansari, A. E. Cetin and S. H. Lee, "Subband Coding of Images Using Nonrectangular Filter Banks," *Proc. SPIE Vol. 974, Applications of Digital Image Processing,* pp. 315-323, 1988.

41 D.E. Dudgeon and R.M. Mersereau, *Multidimensional Digital Signal Processing,*

Prentice-Hall, Englwood Cliffs, NJ, 1984.

42 D. J. Le Gall and A. Tabatabai, "Sub-band Coding of Digital Images Using Short Kernel Filters and Arithmetic Coding Techniques," *Proc. IEEE Int. Conf. on Acoust., Speech, Signal Processing,* April 1988.

43 G. Karlsson and M. Vetterli, "Subband Coding of Video for Packet Networks," *Optical Engineering,* Vol. 27, No. 7, July 1988.

44 S.H Lee and L.T. Wu, "Variable Rate Video Transport in Broadband Packet Networks," *SPIE conf. on Visual Communication and Image Representation,* Nov. 1988, Boston.

45 C. Guillemot and R. Ansari, "Layered Coding Schemes for Video Transmission on ATM Networks", *accepted for publication in Journal on Visual Communication and Image Representation.*

46 M. Nomura, T. Fuji and N. Ohta, "Layered Coding for ATM based Video Distribution Systems," *Signal Processing: Image Communication,* Vol. 3, No. 4, Sep. 1991.

47 Y. Wang and V. Ramamoorthy, "Image Reconstruction from Partial Subband Images and its Application in Packet Video Transmission," *Signal Processing: Image Communication,* Vol. 3, No. 3, June 1991.

48 W-H. Chen and W.K. Pratt, "Scene Adaptive Coder," *IEEE Trans. Comm.* vol. 32, pp.225-232, March 1984.

49 MPEG *Video Committee Draft,* ISO-IEC JTC1/SC2/WG11, MPEG 90/176, Rev. 2, December 18, 1990.

50 R. Ansari and C. Guillemot, "A Hierarchical Scheme for Video Coding at 4 and 9 Mbits/s," *Proceedings EUSIPCO 92,* pp. 227-230.

51 A. N. Akansu and R. A Haddad, "Multiresolution Signal Decomposition, Transforms, Subbands, Wavelets," *Academic Press,* 1992.

52 O. Rioul, "Simple Regularity Criteria for Subdivision Schemes", *SIAM, J. Math. Anal.,* 1991.

53 O. Rioul, "Simple, Optimal Regularity Estimates for Wavelets", *Proceedings EUSIPCO 92,* pp. 937-940.

54 M. Antonini, M. Barlaud and P. Mathieu, "Image Coding Using Wavelet Transform," *IEEE Trans. Acoust., Speech, Signal Proc.,* To appear.

64

Table 7. SNR and Bit Rates for Barbara Image.

DCT Block size	Scaling Factor	Filter #1		Filter #2	
		Bit Rate (Bpp)	SNR (dB)	Bit Rate (Bpp)	SNR (dB)
8x8	1	1.34	33.9	1.28	34.9
	2	1.1	34.0	1.04	34.2
16x16	1	1.1	33.9	1.01	34.2

Table 8. SNR and Bit Rates for Kiel Harbour Image.

DCT Block size	Scaling Factor	Filter #1		Filter #2	
		Bit Rate (Bpp)	SNR (dB)	Bit Rate (Bpp)	SNR (dB)
8x8	1	1.57	31.8	1.57	30.7
	2	1.36	31.5	1.36	30.2
16x16	1	1.37	31.5	1.37	30.4

Chapter 3

WAVELET TRANSFORM AND IMAGE CODING

Marc Antonini, Thierry Gaidon, Pierre Mathieu, Michel Barlaud

I3S Laboratory URA 1376 CNRS
University of Nice-Sophia Antipolis
Bât. 4 SPI - Sophia Antipolis
250, av. A. Einstein
06560 Valbonne FRANCE

Abstract

In many different fields, digitized images are replacing conventional analog images as photograph or X-rays. The volume of data required to describe such images slows down transmission and makes storage prohibitively costly. The information contained in the image must therefore be compressed by extracting only the interesting elements, which are then encoded. These interesting elements are defined according to the post-processing. The quantity of data involved is thus substantially reduced. We propose a new scheme for image compression taking into account the psychovisual effect as well as the statistical redundancies in the image data, enabling bit rate reduction. This new method involves two steps. First, we use a wavelet transform in order to decompose the image at different scales. Second, according to Shannon's rate distortion theory, the wavelet coefficients are vector quantized using lattice vector quantizers. These theoretical results are experimentally checked and some of the coded images are shown in this chapter.

Key words: multiresolution analysis, wavelets, coding, vector quantization, lattice.

1. WAVELET TRANSFORM

1.1. INTRODUCTION

The advantage of numerous transforms for information compression applications is that they project the signal onto a basis of orthogonal functions, i.e., they distribute the signal's energy over a set of decorrelated components. There are a variety of orthogonal transforms, each with specific properties.
The discrete Fourier transform (DFT), the discrete cosine and sine transforms (DCT, DST), the Karhunen-Loeve (KL) transform, and the Haar and Hadamard transforms are the most well known and widely utilized.

The Karhunen-Loeve transform is an optimal transform in that it diagonalizes the covariance matrix [33]. The lack of a rapid algorithm, however, makes the DCT more attractive and in many cases, this latter yields comparable results [3]. The drawback of the DCT is that it does not have a rigorous convolution property [49], and therefore the DFT in its FFT version is often preferred. These transforms do a good job in localizing the energy in the frequency domain, but not in the time domain since they do not admit non-stationary properties.
The Haar and Hadamard transforms, on the other hand, offer good localization in the time domain but not in the frequency domain [2].
The Wigner-Ville transform, which is not orthogonal, can be used for non-stationary signals, but at the cost of significant drawbacks. The fact that it is bilinear makes matrix inversion a difficult task [57], and since it is not a projection onto a basis of functions, the number of points increases as N^2. N being the number of signal points.
Lastly, the Gabor functions, which are well localized in both the space and frequency domains, do not have associated digital filters as a principle (unless the continuous functions are sampled). Implementation of this transform is thus cumbersome [26].

The *wavelet transform* defined by Y. Meyer et J. Lemarié [36] admits non-stationary signals, offers "good" localization in both the space and frequency domains, and can be implemented in a fast algorithm. These properties, and the fact that it is particularly suitable to image signals and takes human vision mechanisms into account [40], make the wavelet transform an ideal candidate for image signal processing. It will be the topic of this chapter. A complete survey of pyramidal representations of digital images is given in [8].

In paragraphs 1.2 and 1.3, wavelets and multiresolution analysis are defined. In paragraph 1.4, the relationship between multiresolution analysis and digital filter banks is discussed. In paragraph 1.5, the quality criteria of wavelets are investigated, first in terms of their suitability to image processing, then as a means of classifying various wavelet bases. A biorthogonal version of the wavelet transform adapted to image signals and implemented using FIR filters is developed in paragraphs 1.6. Extension to the two-dimensional case using separable 1D filters or 2D filters is presented in paragraph 1.7. Paragraph 1.8 is devoted to the statistical properties of wavelet coefficient sub-images.

1.2. DEFINITION OF WAVELETS [1]

Wavelets are functions generated from one single function, the *mother wavelet* ψ by dilation and translation. Grossmann and Morlet [31] introduced this function ψ which dilated by a *scaling factor a* and translated by b enables the analysis, processing, and synthesis of a signal.

$$\psi_{a,b}(x) = |a|^{-1/2} \psi\left(\frac{x-b}{a}\right) \qquad (a,b) \in \mathbf{R}^2, \ a \neq 0 \tag{1-1}$$

We assume x is a one-dimensionnal variable. The mother wavelet ψ has to satisfy the following admissibility condition.

$$\int \frac{|\Psi(\omega)|^2}{|\omega|} d\omega < \infty \tag{1-2}$$

where Ψ denotes the Fourier transform of ψ. Moreover if ψ has sufficient decay, then (1-2) is equivalent to

$$\int_{-\infty}^{+\infty} \psi(x)\, dx = 0 \tag{1-3}$$

which means that the wavelet ψ exhibits at least a few oscillations, and that there is a large choice of functions for ψ [32].

The basic idea of the wavelet transform is to represent any arbitrary function f as a superposition of wavelets. This function f can then be decomposed at different scale or resolution levels. One way to achieve such a decomposition involves writing f as an integral of $\psi_{a,b}$ over a and b using appropriate weighting coefficients [31]. In practice, however, it is preferable to express f as a discrete sum rather than as an integral. The coefficients a and b are thus discretized such that:

$$a = a_0^m \text{ and } b = nb_0 a_0^m \text{ with } (m,n) \in \mathbf{Z}^2 \text{ and } a_0 > 1, \ b_0 > 0 \text{ fixed.}$$

The wavelet is then defined as follows:

$$\psi_{m,n}(x) = \psi_{a_0^m, nb_0 a_0^m}(x) = a_0^{-m/2} \psi\left(a_0^{-m} x - nb_0\right) \tag{1-4}$$

and the wavelet decomposition of f becomes [22], [24]

---

[1] In this Chapter **R** stands for the set of real numbers, **Z** for the set of integer numbers, and **N** for the natural one.

$$f = \sum_{m,n} c_{m,n}(f)\, \psi_{m,n} \tag{1-5}$$

For large, positive values of m ($a>1$), the ψ function is highly dilated and large values for the translation step b are well adapted to this dilation. This corresponds to low frequency or narrow-band wavelets. For large negative values of m ($a<1$), the ψ function is highly concentrated and the translation step b takes small values. These functions correspond to high frequency or wide-band wavelets.

Y. Meyer [43] showed that there are ψ functions, for $a_0 = 2$ and $b_0 = 1$, such that the functions $\psi_{m,n}(x)$ make up an orthonormal basis belonging to $L^2(R)$, where

$$\psi_{m,n}(x) = 2^{-m/2}\, \psi\!\left(2^{-m}x - n\right) \qquad (m,n) \in Z^2 \tag{1-6}$$

The wavelet coefficients $c_{m,n}(f)$ are determined using the following relation:

$$c_{m,n}(f) = \langle f, \psi_{m,n} \rangle = \int f(x)\, \overline{\psi}_{m,n}(x)\, dx \tag{1-7}$$

The oldest known basis of this type was constructed by Haar. In this case, the function $\psi(x)$ is equal to 1 over the interval [0,1/2[, -1 over [1/2,1[and 0 elsewhere. Different bases corresponding to more regular wavelets were later constructed by Stromberg [50], Meyer [43], Lemarié [35], Battle [11], and Daubechies [25].

The existence of orthonormal wavelet bases is conditioned by the following regularity property [44], [24]:

$|\Psi(\omega)|$ must decrease more rapidly than $C(1+|\omega|)^{-\varepsilon-0.5}$ for $\omega \to \infty$ and for $\varepsilon > 0$
Where C is a constant.

Wavelets, which exhibit this regularity property, necessarily verify:

$$\int x^r \psi(x)\, dx = 0 \tag{1-8}$$

This equation determines the number of vanishing moments of ψ [44], and thus enables evaluation of the oscillations of the wavelet ψ.

1.3. MULTIRESOLUTION ANALYSIS - SCALING FUNCTION

In all of the preceeding examples, the orthonormal wavelet bases correspond to *multiresolution analysis*. The concept of multiresolution analysis, introduced by S. Mallat [37], is a mathematical tool particularly well adapted to the use of wavelet bases in image analysis. A brief summary of multiresolution analysis is given below. For further details, the reader may consult [37], [44].

Multiresolution analysis consists of a series of spaces $V_m \subset L^2(\mathbf{R})$, $m \in \mathbf{Z}$

$$... \subset V_2 \subset V_1 \subset V_0 \subset V_{-1} \subset V_{-2} \subset ...$$

with $\overline{\bigcup_{m \in \mathbf{Z}} V_m} = L^2(\mathbf{R})$ $\quad and \quad$ $\bigcap_{m \in \mathbf{Z}} V_m = \{0\}$

These spaces describe successive approximation spaces: for a given function f, the successive projections $Proj_{V_m}(f)$ on the spaces V_m give approximations of f with resolution 2^{-m}

To define multiresolution analysis, a *scaling function* ϕ is introduced. As in the case of wavelets, dilated and translated versions of the scaling function are also introduced

$$\phi_{m,n}(x) = 2^{-m/2} \phi\left(2^{-m}x - n\right) \qquad (m,n) \in \mathbf{Z}^2 \tag{1-9}$$

It then follows that, for any m, the $\left(\phi_{m,n}\right)_{n \in \mathbf{Z}}$ constitute an orthonormal basis for V_m.
The approximation of f with resolution 2^{-m} is written

$$Proj_{V_m}(f) = \sum_{n \in \mathbf{Z}} \langle f, \phi_{m,n} \rangle \phi_{m,n} \tag{1-10}$$

It has been shown that these applications are continuous projectors which degrade the information contained in f as m increases, and which provide an increasingly accurate approximation of f as m decreases. In addition, the ϕ functions necessarily satisfy an equation of the following type [24]:

$$\phi(x) = \sum_{n \in \mathbf{Z}} \alpha_n \phi(2x - n) \tag{1-11}$$

where the orthogonality of the $\phi_{0,k}$ implies that

$$\sum_n \alpha_n \alpha_{n+2k} = 2\delta_{k,0} \tag{1-12}$$

δ is the Kronecker symbol : $\delta_{n,n'} = 1$ if $n = n'$ and 0 elsewhere.
The orthonormal wavelet basis associated with this multiresolution analysis is thus defined below

$$\psi(x) = \sum_{n \in \mathbf{Z}} (-1)^n \alpha_{-n+1} \phi(2x - n) \tag{1-13}$$

where the coefficients α_n are given by equation (1-11), and are considered real here, and hence the functions ϕ and ψ are real.

In multiresolution analysis, i.e., with a family of spaces V_m and a function ϕ satisfying the conditions stated in [24], we define a space W_m which is the orthogonal complement of V_m in space V_{m-1}:

$$V_{m-1} = V_m \oplus W_m \quad , \quad V_m \perp W_m \tag{1-14}$$

For a given m, the functions $\psi_{m,n}$ constitute an orthonormal basis in W_m The coefficients $\left\{ c_{m,n}(f) = \langle f, \psi_{m,n} \rangle ; n \in Z \right\}$ characterize the difference between two approximations of f at different resolutions, i.e., the information lost when going from an approximation of resolution 2^{-m+1} to a coarser approximation of resolution 2^{-m}. It can then be shown that [44]:

$$Proj_{V_{m-1}}(f) = Proj_{V_m}(f) + \sum_{n \in Z} \langle f, \psi_{m,n} \rangle \psi_{m,n} \tag{1-15}$$

1.4. MULTIRESOLUTION ANALYSIS AND DIGITAL FILTER BANKS

In signal processing, and particularly in image processing, the signals used are sampled. The samples of the original signals are considered to be the approximation coefficients, which will be called $s_0(n)$, whose resolution is determined by the sampling rate.

These sampled signals are processed in the <u>spatial and frequency domains</u> using digital filters which generate orthogonal multiresolution analysis $(V_m)_{m \in Z}$ of $L^2(R)$. However, not all digital filters give rise to multiresolution analysis. In the following sections, we state, without demonstration, the relationships between sampled filters and the continuous basis functions ϕ and ψ (see § 1.4.1.1.), as well as the sufficient condition for the construction of orthogonal multiresolution analysis (see § 1.4.1.2.). The curious reader is invited to consult [37], [24], [17] for the theoretical groundwork, which is beyond the scope of the present chapter.

The rest of this paragraph will focus on orthonormal wavelet bases.

1.4.1. A brief review of orthogonal multiresolution analysis

1.4.1.1. <u>Relationships between digital filters and wavelets</u>

Let $\left(V_m\right)_{m \in Z}$ be a multiresolution analysis in $L^2(R)$ and $\phi_m(x)$ an <u>orthonormal basis</u> in V_m. In what follows, we will consider the functions ϕ and ψ to be real.

The function $\phi_{m+1,n}(x) \in V_{m+1} \subset V_m$ can be decomposed onto the basis of functions $\phi_m(x)$ through the following relation:

$$\phi_{m+1,n}(x) = \sum_{k=-\infty}^{+\infty} \langle \phi_{m+1,n}, \phi_{m,k} \rangle \phi_{m,k}(x) \tag{1-16}$$

where

$$\langle\phi_{m+1,n},\phi_{m,k}\rangle = \int\limits_{-\infty}^{-\infty} 2^{-(m+1)/2}\, \phi\!\left(2^{-(m+1)}x-n\right) 2^{-m/2}\, \phi\!\left(2^{-m}x-k\right) dx \tag{1-17}$$

If we perform the transformation $2^{-(m+1)}x-n = 2^{-1}y$ the following equation is obtained:

$$\langle\phi_{m+1,n},\phi_{m,k}\rangle = \int\limits_{-\infty}^{+\infty} 2^{-1/2}\, \phi\!\left(2^{-1}y\right)\phi(y-(k-2n))\, dy \tag{1-18}$$

It is then possible to write equation (1-19) which means that the scalar product is independent of the resolution m:

$$\langle\phi_{m+1,n},\phi_{m,k}\rangle = \langle\phi_{1,0},\phi_{0,k-2n}\rangle \tag{1-19}$$

If we project a function f onto the basis functions $\phi_{m+1,n}(x)$ from equations (1-16) and (1-19), we obtain:

$$\langle f,\phi_{m+1,n}\rangle = \sum_{k=-\infty}^{+\infty}\langle f,\phi_{m,k}\rangle\langle\phi_{1,0},\phi_{0,k-2n}\rangle \tag{1-20}$$

We can then set

$$h(n) = \langle\phi_{1,0},\phi_{0,n}\rangle = \int\limits_{-\infty}^{+\infty} 2^{-1/2}\, \phi\!\left(\frac{1}{2}x\right)\phi(x-n)\, dx \tag{1-21}$$

which is the impulse response of a low-pass filter. Daubechies showed in [24] that these coefficients $h(n)$ correspond (with the normalization factor $2^{-1/2}$) exactly to the coefficients α_n in equation (1-11).

Let $s_{m+1}(n) = \langle f,\phi_{m+1,n}\rangle$ be the sampled signal corresponding to the projection of f on the basis formed by $\phi_{m+1,n}(x)$ and $s_m(n) = \langle f,\phi_{m,n}\rangle$ the sampled signal corresponding to the projection of f on the basis formed by $\phi_{m,n}(x)$. For filters with an impulse response $h(n) = h(-n)$, from equations (1-20) and (1-21) we can write:

$$s_{m+1}(n) = \langle f,\phi_{m+1,n}\rangle = \sum_{k=-\infty}^{+\infty} h(2n-k)\langle f,\phi_{m,k}\rangle = \sum_{k=-\infty}^{+\infty} h(2n-k)s_m(k) \tag{1-22}$$

This equation shows that the coefficients $s_{m+1}(n)$ can be computed by convolving $h(n)$ with $s_m(n)$. The s_j, for any $j>0$, are then obtained by iterating this convolution. This *pyramidal transform*, as introduced by S. Mallat, will be discussed in paragraph 1.4.1.3.

Combining (1-19), (1-21), and (1-'5), we find the following equation for $m=0$, which expresses the relationship between the digital filter h and the scaling function ϕ

$$\phi\left(\frac{x}{2}\right) = \sqrt{2} \sum_{k=-\infty}^{+\infty} h(k)\,\phi(x-k) \tag{1-23}$$

This convolution product is equivalent to a function $H(\omega) \in L^2\big([0, 2\pi[\big)$ in the Fourier domain such that:

$$\Phi(2\omega) = H(\omega)\,\Phi(\omega) = \prod_{n=0}^{+\infty} H\left(\frac{\omega}{2^n}\right) \tag{1-24}$$

where

$$H(\omega) = \frac{1}{\sqrt{2}} \sum_n h(n) e^{-in\omega} \tag{1-25}$$

The orthonormality condition on the $\phi_{m,n}$ functions is equivalent to

$$\sum_{n=-\infty}^{+\infty} \left| \Phi(\omega + 2n\pi) \right|^2 = 1 \tag{1-26}$$

Combining (1-24) and (1-26), we obtain the exact reconstruction equation (A-12) of the conjugate quadrature filters (CQFs) with a factor 2 (see appendix A.1)

$$\left| H(\omega) \right|^2 + \left| H(\omega + \pi) \right|^2 = 1 \tag{1-27}$$

This equation shows that H is necessarily a function whose modulus is bounded between 0 and 1. Let $H(0)=1$ and $H(\pi)=0$. The filter $\sqrt{2}\,H(\omega)$ thus generates a bank of two conjugate quadrature filters enabling exact reconstruction as defined in appendix A.2.

The relationships developed for a low-pass filter $h(n)$ are also valid for the corresponding high-pass filter $g(n)$, defined in the filter bank presented in appendix. The following expression can be employed [24]

$$g(n) = \int_{-\infty}^{+\infty} 2^{-1/2}\, \psi\left(\frac{1}{2}x\right) \phi(x-n)\, dx = (-1)^n h(-n+1) \tag{1-28}$$

leading to

$$\psi\left(\frac{x}{2}\right) = \sqrt{2} \sum_{n=-\infty}^{+\infty} (-1)^n h(-n+1)\,\phi(x-n) \tag{1-29}$$

and

$$\Psi(2\omega) = \overline{H}(\omega + \pi)e^{-i\omega}\,\Phi(\omega) \tag{1-30}$$

The coefficients $h(n)$ and $g(n)$ are chosen according to the desired multiresolution analysis. However, $h(n)$ must satisfy the following normalization condition to ensure convergence of (1-24) [24].

$$\sum_n h(n) = \sqrt{2} \tag{1-31}$$

and expression (1-5) is verified if [24]

$$\sum_n g(n) = 0 \tag{1-32}$$

Figure 1.1. depicts a simplified filter bank structure showing the relationship between orthonormal wavelets and digital filters. $\tilde{H}(\omega) = \overline{H}(\omega)$ and $\tilde{G}(\omega) = \overline{G}(\omega)$.

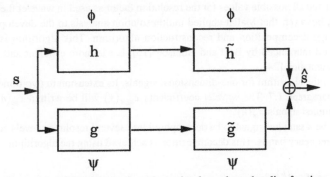

Figure 1.1.: Digital filter banks and associated wavelet and scaling function.

1.4.1.2. Condition for the construction of an orthogonal multiresolution analysis

A. Cohen recently showed (1989) that not all quadrature mirror filters enable multiresolution analysis, and determined a necessary and sufficient condition on filter h for multiresolution analysis [17]. This major mathematical contribution goes beyond the scope of the present chapter, which will focus simply on the first theorem demonstrated by Mallat [38], which enables construction of orthogonal multiresolution analyses from the filter function h, i.e., from the CQFs.

<u>Theorem 1</u> : Let $H(\omega)$ be a function of class C^1 and 2π periodic verifying:

$$H(0)=1,$$

$$|H(\omega)|^2 + |H(\omega+\pi)|^2 = 1,$$

$$\exists \rho > 0 \text{ such that } \forall \omega \in \mathbf{R}, |\omega| < \frac{\pi}{2} \Rightarrow |H(\omega)| > \rho$$

then the function ϕ defined by the recursive formula (1-24) leads to orthogonal multiresolution analysis. Inversely, multiresolution analysis always results in a pair of quadrature conjugate filters [45] (see appendix A).

1.4.1.3. The orthogonal multiresolution analysis algorithm: Mallat's algorithm

Relationship (1-21) between quadrature mirror filters and orthogonal wavelets, presented in paragraph 1.4.1.1. was established in 1986 by S. Mallat [37], who showed that the wavelet coefficients defined by $c_{m,n}(f) = \langle f, \psi_{m,n} \rangle$ can be computed from a <u>pyramidal transform</u> implemented using digital filters. This orthogonal multiresolution analysis is characterized by a resolution factor of 2 between two consecutive scale levels, and is thus called *dyadic* multiresolution analysis. As pointed out by Daubechies [24], this resolution factor can take other values, but not arbitrarily. Nonetheless, the value 2 and now $\sqrt{2}$ [28] are the only values actually studied and utilized. Meyer, from a more general standpoint, evaluated all possible values for the resolution factor as used in wavelet theory [43].

It is to be noted, however, that Mallat applied multiresolution analysis to the development of a two-dimensional image decomposition and reconstruction algorithm. This algorithm is similar to the Laplacian pyramid introduced by Burt and Adelson [15], but is more efficient and enables a finer choice of edge orientation [24].

We now focus on the algorithm for one-dimensional signals. Its extension to two-dimensional signals is described in paragraph 1.7. The wavelet coefficients $c_{m,n}(f)$ will be written $c_m(n)$, since we are working with sampled signals $s_m(n)$.

Let $s_0(n) \in l^2(\mathbf{Z})$ be a sampled signal to be decomposed into several resolution levels corresponding to different space-frequency bands. This decomposition is achieved using the algorithm:

$$s_m(n) = \sum_k h(2n-k)s_{m-1}(k)$$

$$c_m(n) = \sum_k g(2n-k)s_{m-1}(k)$$

(1-33)

The signal $s_m(n)$ is an approximation of signal $s_{m-1}(n)$ at resolution 2^{-m}. The wavelet coefficients $c_m(n)$ represent the information lost when going from an approximation of the signal s with resolution 2^{-m+1} to a coarser approximation of s with a resolution 2^{-m}. According to Shannon's theorem, these signals are undersampled by a factor 2 (see appendix A). This analysis algorithm is identical to the conjugate quadrature filter (CQF) decomposition algorithm. $h(n)$ is the low-pass filter and $g(n)$ is the high-pass filter. These two filters are related by:

$$g(n) = (-1)^n h(-n+1)$$

(1-34)

Since these filters are associated with an orthonormal wavelet basis, they ensure the exact reconstruction of the signal $s_{m-1}(n)$. The reconstruction formula is as follows:

$$s_{m-1}(k) = \sum_n h(2n-k)s_m(n) + \sum_n g(2n-k)c_m(n) \tag{1-35}$$

The basic principle of multiresolution analysis involves decomposing a signal $s_0(n)$ into two "subsignals" $s_1(n)$ and $c_1(n)$. This operation can then be repeated on signal $s_1(n)$ and so on up to resolution 2^{-j}. In this case, the signal set $c_1 \cup c_2 \cup c_3 \cup \ldots \cup c_j \cup s_j$ provides a lossless representation of s_0, and hence enables the exact reconstruction of this signal.

Figure 1.2 gives the decomposition of signal s_0 up to resolution 2^{-j}. In figure 1.3, we recall the filter bank structure used for the analysis and reconstruction of the signal using Mallat's algorithm. This structure is identical to that given in the appendix A for QMF or CQF filters.

To better illustrate the principle shown in figure 1.2, we give, in figure 1.4, an example of multiresolution decomposition of a one-dimensional sampled signal.

1.4.2. Daubechies' compactly supported wavelets

In 1987, Daubechies proposed a method for building compactly supported wavelet bases by imposing sufficient conditions on the filter $H(\omega)$. These conditions, expressed algebraically, are such that $H(\omega)$ generates multiresolution analysis, but they are not based on an *a priori* definition of analysis spaces $(V_m)_{m \in \mathbb{Z}}$.

The following paragraph briefly recalls these conditions.

In order to achieve compactly supported wavelets, it is necessary to fix the number of coefficients of filter $h(n)$. $H(\omega)$, which is the Fourier transform of $h(n)$, to within a certain factor (eq. (1-25)), is built by setting

$$H(\omega) = \left[\frac{1}{2}\left(1 + e^{i\omega}\right) \right]^N Q\left(e^{i\omega}\right) \tag{1-36}$$

where Q is a trigonometric polynomial of the form

$$Q\left(e^{i\omega}\right) = \sum_{n=0}^{N-1} q(n)e^{in\omega} \qquad q(0) \neq 0 \tag{1-37}$$

From equation (1-37), we can write the following

$$|H(\omega)|^2 = \left(\cos^2\left(\frac{\omega}{2}\right)\right)^N \left|Q\left(e^{i\omega}\right)\right|^2 \tag{1-38}$$

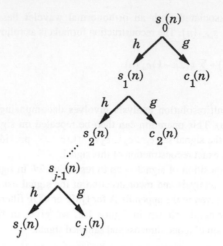

Figure 1.2. : Decomposition of a signal $s_0(n)$ from scale 2^0 to scale 2^j

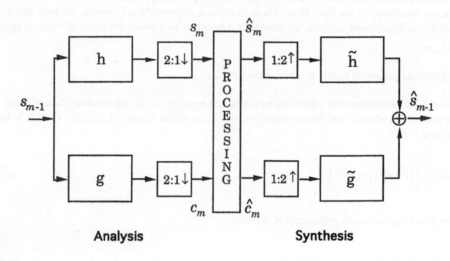

Analysis **Synthesis**

| X | : Convolve with filter X |

| 2:1↓ | : Keep one sample out of two (undersampling) |

| 1:2↑ | : Put a zero between each sample (oversampling) |

Figure 1.3. : Mallat's algorithm: general signal analysis and reconstruction scheme.

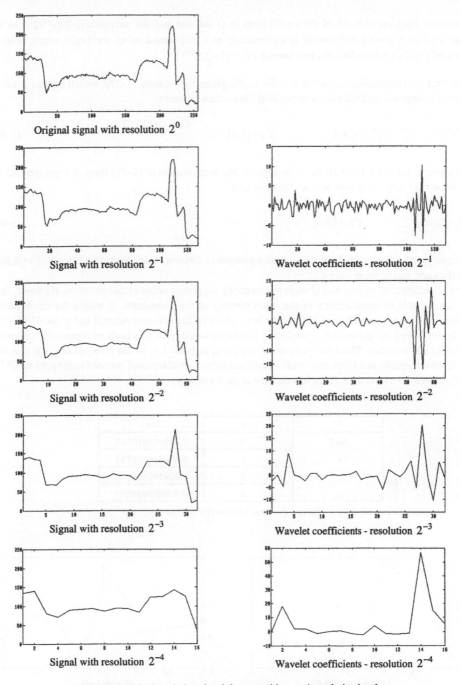

Figure 1.4. : Multiresolution signal decomposition on 4 resolution levels.

Since the $h(n)$ are real, all of the coefficients in Q are real and the polynomial $\left|Q\left(e^{i\omega}\right)\right|^2$ can be expressed as a $\cos(\omega)$ polynomial or equivalently as a polynomial using $\sin^2(\omega/2)$ terms. Setting $y = \cos^2(\omega/2)$, we introduce the polynomial $P(1-y) = \left|Q\left(e^{i\omega}\right)\right|^2$.

The exact reconstruction equation for the CQFs given by formula (1-27), which is also a basic condition imposed by Daubechies on the $h(n)$ filters, then becomes:

$$y^N P(1-y) + (1-y)^N P(y) = 1 \qquad \forall y \in [0,1] \tag{1-39}$$

As a result, for every filter $H(\omega)$ of the type (1-36), and solution to (1-27) there is a polynomial P, solution to (1-39) which also satisfies the condition

$$P(y) \geq 0 \qquad \forall y \in [0,1] \tag{1-40}$$

Inversely, any polynomial P satisfying both equations (1-39) and (1-40) leads to a filter $h(n)$ with real coefficients, and solution to (1-27).

Thus, the choice of polynomial Q leads to compactly supported orthonormal wavelets. However, this choice is made in consideration of the zero-crossings of the polynomial Q within the circle unity, which corresponds to a minimal-phase filter $h(n)$, which is thus non-symmetrical and of the CQF type. In the following, we present an example of a filter $h(n)$ which leads to a basis of compactly supported orthonormal wavelets. The filter parameters are given in table 1.1. In the present case, we selected $N=2$, which results in a filter $h(n)$ with 4 coefficients and an associated wavelet belonging to $C^{0.5-\varepsilon}$ [24]. Figure 1.5 shows the scaling function ϕ and the wavelet ψ.

	n	$h(n)$
$N=2$	0	0,482962913145
	1	0,836516303738
	2	0,224143868042
	3	-0,129409522551

Table 1.1.: Filter $h(n)$ coefficients for $N=2$

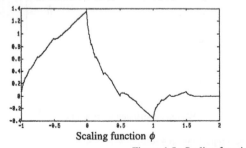
Scaling function ϕ

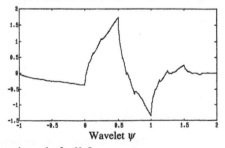
Wavelet ψ

Figure 1.5.: Scaling function and wavelet for $N=2$.

1.5. QUALITY CRITERIA FOR WAVELETS USED IN IMAGE PROCESSING

This section is devoted to a classification of wavelet bases in terms of their performance in image compression applications. A major problem involves defining the criteria for this classification. Below, we present a number of wavelet properties which appear to us essential for the processing of image signals.

1.5.1. Wavelet regularity

The *regularity* of the mother wavelet is important and appears to be closely related to the regularity of the signal to be processed. Since images are generally "smooth" to the eye, with the exception of occasional edges, it is appropriate to use regular wavelets. Indeed, there is a trade-off between wavelet regularity and the visual effect on the processed image.

For I. Daubechies [24], the wavelet is considered regular if the functions ϕ and ψ belong to class C^α where $\alpha \in Z^+$. For a given function f

$$f \in C^\alpha \Leftrightarrow \int |F(\omega)| (1+|\omega|)^{1+\alpha} \, d\omega < \infty \tag{1-41}$$

Thus f is int(α) times continuously differentiable (int(α) being the whole part of α). However, it must be noted that, for $\alpha = k \in N$, condition (1-41) implies that $f \in C^k$, but this latter is not necessary [24].

A numerical method for measuring this regularity proposed by [28] involves the systematic search for an exponent v such that

$$|\Phi(\omega)| \le (1+|\omega|)^v \quad \text{for } |\omega| \text{ greater than a certain value.}$$

This means that ϕ is at least in class C^{-v-1}.

Note that there exist other methods to estimate the regularity of a wavelet. See for example the recent works of Rioul [46,47].

1.5.2. Number of vanishing moments

Another important criterion is the number N of *vanishing moments* of the wavelet ψ, i.e., its oscillatory character. The number of vanishing moments was defined in paragraph 1.2 (formula (1-8)); it is recalled below:

$$\int x^n \psi(x) \, dx = 0 \quad \forall n = 0,1,\ldots,N-1$$

Since ψ is a wavelet, N is greater than or equal to 1 and the property (1-3) is verified. Actually, this number is related to the wavelet's regularity; any r-regular multiresolution analysis generates a wavelet ψ with $r+1$ vanishing moments. Furthermore, all derivatives of $\Psi(\omega)$ up to the order $N-1$ are zero at the point $\omega = 0$.

In practice, a wavelet with N vanishing moments enables the cancellation of all wavelet coefficients of a polynomial signal whose degree is less than N. Thus, if f is a polynomial signal of degree less than N, on the support of $\psi_{m,n}$, then

$$c_{m,n}(f) = \langle f, \psi_{m,n} \rangle = 0$$

This result is quite significant for image coding applications because it enables high compression rates (many wavelet coefficients are zero or negligible).

1.5.3. Spatial characterization of the scaling function

To determine how $\phi(x)$ evolves with respect to x, we introduce the criterion m_k which allows us to characterize the scaling function ϕ spatially. To define m_k, we introduce the function $p(x)$ such that:

$$p(x) = \frac{|\phi(x)|^2}{\int |\phi(x)|^2 \, dx} \qquad \text{with} \quad p(x) \geq 0 \quad \text{and} \quad \int p(x) \, dx = 1 \qquad (1\text{-}42)$$

m_k is then defined by $m_k = \int (x - m_1)^k \, p(x) \, dx$ $\qquad\qquad\qquad\qquad\qquad$ (1-43)

with $m_1 = \int x \, p(x) \, dx$

Thus m_1 is equal to the mathematical expectation of $p(x)$ and m_2 corresponds to the "spatial variance" of the scaling function ϕ. m_2 allows us to determine the energy concentration of ϕ and provides information on the spatial length or localization of ϕ (cf. table 1.6).
These criteria also apply to the wavelet ψ.

1.5.4. Characterization of the associated filters

To avoid distortion in image processing, the filter $H(\omega)$ associated with the scaling function ϕ must be linear phase or ideally *zero phase*. Indeed, non-linear phase filters degrade edges and are more difficult to implement than linear phase filters. Furthermore, figure 1.6. shows the effects of the phase in image processing. In addition, the number of elements making up the impulse response of $h(n)$ must be small in order to limit the number of convolution operations to be performed in the analysis/reconstruction algorithm. It corresponds to wavelets whose support is compact (making the wavelet well localized).

The following paragraphs deal with the relative importance of these properties. However, it is not currently known which of these properties will prove determinant, although it can be seen that several are mutually exclusive: linear phase filter and compactly supported wavelet, short filter and wavelet exhibiting good regularity, etc...

Lena
        Cameraman

(a) - Lena image with cameraman phase   (b) - Cameraman image with lena phase

Figure 1.6.: Sensibility of the human eyes with the image phase.
(a) - Modulus of lena image with cameraman phase (b) - Modulus of cameraman image with lena phase.

1.6. BIORTHOGONAL WAVELETS

1.6.1. Why use biorthogonal wavelets ?

The filtering technique based on the use of filter banks as described previously is well known in the signal processing community, and was developed by Smith and Barnwell [48] and Vetterli [54] shortly before multiresolution analysis was proposed in mathematics. Under certain conditions, this technique can generate a highly useful orthogonal multiresolution analysis when the signal's characteristics are sought at different scales. However, it is impractical in image processing because the associated filters are nonlinear phase. Indeed, since images are, for the most part, smooth to the eye, it would seem appropriate to use exact reconstruction filters corresponding to an orthonormal wavelet basis with a mother wavelet exhibiting good regularity. In addition, in order to perform rapid convolutions, the FIR filters used must be short. On the other hand, these filters should be linear phase (and even zero phase) as explained in paragraph 1.5.4. Unfortunately, not all of these conditions can be satisfied simultaneously since there are no orthonormal linear phase FIR filters enabling exact reconstruction [48], regardless of regularity. The only symmetric exact reconstruction filters are those which correspond to the Haar basis.

Nonetheless, the highly important linear phase constraint corresponding to symmetrical wavelets can be maintained by relaxing the orthonormality constraint and by using *biorthogonal bases*. We can then construct filters asociated with wavelets exhibiting a high degree of regularity.

Biorthogonal bases of wavelets with good regularity were recently developed simultaneously, yet independently by Cohen, Daubechies, and Feauveau [18] and by Herley and Vetterli [55]. Cohen proposed a detailed mathematical investigation with proofs that, under certain conditions, wavelets make up numerically stable bases (Riesz bases) and a discussion on the necessary and sufficient conditions for regularity. In [28], Feauveau offered a new approach for the construction of biorthogonal bases starting from multiresolution spaces rather than from filters. This approach is more similar to that presented by Mallat in his overview article [37].

1.6.2. Non-orthogonal multiresolution analysis

1.6.2.1. Approximation spaces and biorthogonality

Biorthogonal wavelet bases are a generalization of orthogonal wavelet bases. Indeed, in the biorthogonal case, there are in fact two *dual bases* $\psi_{m,n}(x)$ and $\tilde{\psi}_{m,n}(x)$, each achieved by dilations and translations of a single function $\psi(x)$ or $\tilde{\psi}(x)$. P. Tchamitchian [51] was the first to build pairs of dual non-orthonormal bases, in 1987.

Multiresolution analysis based on biorthogonal wavelet bases is more complex to implement than that using orthonormal bases. Basically, in the biorthogonal case, there are two levels of approximation spaces $V_m \subset L^2(\mathbf{R})$ and $\tilde{V}_m \subset L^2(\mathbf{R})$, $m \in \mathbf{Z}$, such that

$$...\subset V_2 \subset V_1 \subset V_0 \subset V_{-1} \subset V_{-2} \subset...$$
$$...\subset \tilde{V}_2 \subset \tilde{V}_1 \subset \tilde{V}_0 \subset \tilde{V}_{-1} \subset \tilde{V}_{-2} \subset...$$

For any m, the functions $\phi_{m,n}$ and $\tilde{\phi}_{m,n}$ are non-orthogonal bases in the spaces V_m and $\tilde{V}_m$ respectively.

As in the orthonormal case, we introduce space W_m which is the complement to V_m in space V_{m-1}. However, W_m is not the orthogonal complement to V_m. By the same token, we introduce space $\tilde{W}_m$ which is the non-orthogonal complement to $\tilde{V}_m$. This translates as follows:

$$
\begin{aligned}
V_{m-1} &= V_m \oplus W_m \quad , \quad V_m \perp W_m \\
\tilde{V}_{m-1} &= \tilde{V}_m \oplus \tilde{W}_m \quad , \quad \tilde{V}_m \perp \tilde{W}_m
\end{aligned}
\tag{1-44}
$$

The following property also exists [18]

$$
\tilde{W}_m \perp V_m \quad \text{and} \quad W_m \perp \tilde{V}_m
\tag{1-45}
$$

For a given m, the functions $\psi_{m,n}$ and $\tilde{\psi}_{m,n}$ are non-orthogonal bases in spaces W_m et $\tilde{W}_m$.

It is noted that the conditions imposed on the approximation spaces in order to achieve biorthogonality result in the following for the basis functions

$$
\left\langle \psi_{m,n}, \tilde{\psi}_{m',n'} \right\rangle = \delta_{m,m'}\, \delta_{n,n'}
\tag{1-46}
$$

Thus the decomposition of a signal f onto biorthogonal wavelets bases is expressed as follows [18]

$$
f = \sum_{m,n} \left\langle f, \psi_{m,n} \right\rangle \tilde{\psi}_{m,n}
\tag{1-47}
$$

where $c_{m,n}(f) = \left\langle f, \psi_{m,n} \right\rangle$ are the wavelet coefficients as defined in paragraph 1.1. and $\tilde{\psi}_{m,n}$ is the dual basis to $\psi_{m,n}$. Therefore the function $\psi(x)$ is used for analysis of the signal while the dual function $\tilde{\psi}(x)$ is used for synthesis. Note that it is legitimate to call the $\tilde{\psi}(x)$ function a wavelet because we can write

$$
f = \sum_{m,n} \left\langle f, \tilde{\psi}_{m,n} \right\rangle \psi_{m,n}
\tag{1-48}
$$

As in the orthogonal case, we define here continuous projectors $Proj_{\tilde{V}_m}(f)$ such that

$$
Proj_{\tilde{V}_m}(f) = \sum_{n \in Z} \left\langle f, \phi_{m,n} \right\rangle \tilde{\phi}_{m,n}
\tag{1-49}
$$

Functions $\phi(x)$ and $\tilde{\phi}(x)$ thus define the concept of multiresolution analysis. Here, the wavelet's job is to extract the details lost between two consecutive scales. It has been shown that [28]:

$$Proj_{\tilde{V}_{m-1}}(f) = Proj_{\tilde{V}_m}(f) + \sum_{n \in Z} \langle f, \psi_{m,n} \rangle \tilde{\psi}_{m,n} \qquad (1\text{-}50)$$

1.6.2.2. Relationship between digital filters and biorthogonal wavelets

As in the orthonormal case, we introduce the filters h, $\tilde{h}$, and g, $\tilde{g}$, which are related respectively to the scaling functions ϕ, $\tilde{\phi}$ and to the wavelets ψ, $\tilde{\psi}$ as expressed in paragraph 1.4.1.1. We recal these relationships in the biorthogonal case. As presented in figure 1.7, a pair of filters (h, g) is used for signal analysis and another pair $(\tilde{h}, \tilde{g})$ for signal reconstruction. Since we apply biorthogonal multiresolution analysis, these filters are symmetrical and hence their Fourier transforms as defined in (1-25) are real.

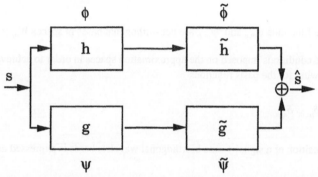

Figure 1.7.: Digital filter banks and the associated wavelets and scaling functions.

By introducing filters h and $\tilde{h}$, we are able to define functions $\phi(x)$ and $\tilde{\phi}(x)$ in terms of their Fourier transform (cf. formula (1-24)) and demonstration paragraph 1.4.1.1). We can thus write:

$$\phi\left(\frac{x}{2}\right) = \sqrt{2} \sum_{n=-\infty}^{+\infty} h(n)\,\phi(x-n)$$
$$\tilde{\phi}\left(\frac{x}{2}\right) = \sqrt{2} \sum_{n=-\infty}^{+\infty} \tilde{h}(n)\,\tilde{\phi}(x-n) \qquad (1\text{-}51)$$

or else

$$\Phi(2\omega) = H(\omega)\,\Phi(\omega) = \prod_{n=0}^{+\infty} H\left(\frac{\omega}{2^n}\right)$$
$$\tilde{\Phi}(2\omega) = \tilde{H}(\omega)\,\tilde{\Phi}(\omega) = \prod_{n=0}^{+\infty} \tilde{H}\left(\frac{\omega}{2^n}\right) \qquad (1\text{-}52)$$

These well defined functions whose modulus is square-integrable are compactly supported if h and $\tilde{h}$ are FIR filters. We are then able to assert that the functions

$$\psi\left(\frac{x}{2}\right) = \sqrt{2} \sum_{n=-\infty}^{+\infty} g(n)\phi(x-n)$$

$$\tilde{\psi}\left(\frac{x}{2}\right) = \sqrt{2} \sum_{n=-\infty}^{+\infty} \tilde{g}(n)\tilde{\phi}(x-n)$$

(1-53)

basically verify equation (1-47) [18]. This means that:

$$\Psi(2\omega) = \tilde{H}(\omega+\pi)e^{-i\omega}\Phi(\omega)$$

$$\tilde{\Psi}(2\omega) = H(\omega+\pi)e^{-i\omega}\tilde{\Phi}(\omega)$$

(1-54)

Finally, if the infinite products defined in (1-52) decay more rapidly than $C(1+|\omega|)^{-\varepsilon-0,5}$ for $\omega \to \infty$ and for all $\varepsilon > 0$ (C is a constant), then ϕ and $\tilde{\phi}$ and by the same token, ψ and $\tilde{\psi}$ are regular [18]. However, it is emphasized that this interpretation is possible only if the infinite products converge, i.e., if

$$H(0) = 1 \quad \text{and} \quad \tilde{H}(0) = 1$$

let

$$\sum_n h(n) = \sqrt{2} \quad \text{and} \quad \sum_n \tilde{h}(n) = \sqrt{2}$$

(1-55)

Formula (1-47) is verified if [18]

$$\sum_n g(n) = 0 \quad \text{and} \quad \sum_n \tilde{g}(n) = 0$$

(1-56)

1.6.3. Nonorthogonal multiresolution analysis algorithm

This part focuses on decomposition and reconstruction algorithms for nonorthogonal multiresolution analysis as well as the exact reconstruction condition, which will be presented as previously in the case of orthogonal multiresolution analysis. Readers interested in the theoretical foundations of the results presented here are invited to consult [18], [28].

Let $s_0(n) \in l^2(Z)$ be a sampled signal to be decomposed at several resolution levels. This decomposition is performed using the following algorithm:

$$s_m(n) = \sum_k h(2n-k)s_{m-1}(k)$$

$$c_m(n) = \sum_k g(2n-k)s_{m-1}(k)$$

(1-57)

which is identical to that used for orthogonal multiresolution analysis, defined in formula (1-33), paragraph 1.4.1.3. The reconstruction algorithm, on the other hand, is expressed

$$s_{m-1}(k) = \sum_n \tilde{h}(2n-k)s_m(n) + \sum_n \tilde{g}(2n-k)c_m(n) \qquad (1\text{-}58)$$

This algorithm has the same structure as the reconstruction algorithm used in the orthogonal case. To ensure exact reconstruction of the signal s, the following relationship must hold between the two filters:

$$\tilde{g}(n) = (-1)^n h(-n+1)$$
$$g(n) = (-1)^n \tilde{h}(-n+1) \qquad (1\text{-}59)$$

and, since $\phi_{0,k}$ and $\tilde{\phi}_{0,k}$ are orthogonal,

$$\sum_n h(n)\tilde{h}(n+2k) = \delta_{k,0} \qquad (1\text{-}60)$$

1.6.4. Choice and computation of the dual filter

In [18], it is shown that the highly regular wavelets ψ and $\tilde{\psi}$ can be achieved with sufficiently long filters. In particular, if the functions ψ and $\tilde{\psi}$ are respectively $(k-1)$ and $(\tilde{k}-1)$ times continuously differentiable, then the trigonometric polynomials $H(\omega)$ et $\tilde{H}(\omega)$ must be divisible by $\left(1+e^{-i\omega}\right)^k$ and $\left(1+e^{-i\omega}\right)^{\tilde{k}}$ respectively. Thus the corresponding filter lengths of h and $\tilde{h}$ must be greater than k and $\tilde{k}$. Furthermore, equations (1-59) show that, if $\tilde{H}(\omega)$ is divisible by $\left(1+e^{-i\omega}\right)^{\tilde{k}}$ then ψ has $\tilde{k}$ vanishing moments.

It is well known and can easily be demonstrated using Taylor expansions that, if ψ has $\tilde{k}$ vanishing moments, then the coefficients $\langle f, \psi_{m,n} \rangle$ represent functions f which are $\tilde{k}$ times differentiable, with a high compression potential (since many coefficients are negligibly small). For a more detailed discussion, see [18].

Figure 1.8 shows how regularity, zero moments, and filter length are related. In this diagram, we have indicated only the link between differentiability of $\tilde{\psi}$ and the number of zero moments of ψ.

A broad range of biorthogonal wavelet bases with regular ψ and $\tilde{\psi}$ wavelets can be constructed. For image compression applications, the regularity of the synthesis functions $\tilde{\psi}_{m,n}$, which is related to the number of zero moments in the analysis functions $\psi_{m,n}$, is more important than the regularity of the $\psi_{m,n}$ or the number of zero moments in the $\tilde{\psi}_{m,n}$. With this consideration, and the limit imposed on filter length, we choose the highest possible values for $\tilde{k}$.

The exact reconstruction condition defined in (1-60) can be written in terms of trigonometric polynomials $H(\omega)$ and $\tilde{H}(\omega)$. Indeed, for symmetric filters, we have

$$H(\omega)\tilde{H}(\omega) + H(\omega+\pi)\tilde{H}(\omega+\pi) = 1 \qquad (1\text{-}61)$$

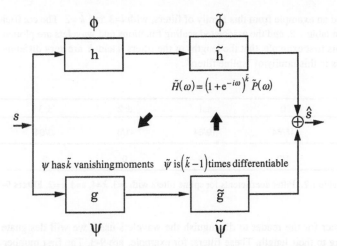

Figure 1.8.: Relationships among regularity, zero moments, and filter length. $\tilde{P}(\omega)$ is a trigonometric polynomial.

This condition, along with the divisibility of $H(\omega)$ and $\tilde{H}(\omega)$ by $\left(1+e^{-i\omega}\right)^k$ and $\left(1+e^{-i\omega}\right)^{\tilde{k}}$ respectively, leads to [24], [18]

$$H(\omega)\tilde{H}(\omega)=\cos^{2l}(\omega/2)\left[\sum_{p=0}^{l-1}\binom{l-1+p}{p}\sin^{2p}(\omega/2)+\sin^{2l}(\omega/2)R(\omega)\right] \tag{1-62}$$

where $R(\omega)$ is an odd polynomial in $\cos(\omega)$ and $2l=k+\tilde{k}$. The symmetry of filters h and $\tilde{h}$ means that $(k+\tilde{k})$ is even. This relationship enables us to compute the dual filter $\tilde{H}(\omega)$ if the filter $H(\omega)$ is imposed.

Numerous examples are possible. The following paragraph focuses on three filters belonging to different families.

1.6.5. Examples of biorthogonal wavelets

1.6.5.1. Spline filters

In (1-62), we can choose $R\equiv 0$ and $\tilde{H}(\omega)=\cos^{\tilde{k}}(\omega/2)\,e^{-i\kappa\omega/2}$ where $\kappa=0$ if $\tilde{k}$ is even and $\kappa=1$ if $\tilde{k}$ is odd. This corresponds to filters called "spline filters" in [18] because the scaling function $\tilde{\phi}$ is a B-spline function. In [55], however, these filters are known as "binomial filters" because the $\tilde{h}$ are simply binomial coefficients. It then follows that

$$H(\omega)=\cos^{2l-\tilde{k}}(\omega/2)\,e^{-i\kappa\omega/2}\left[\sum_{p=0}^{l-1}\binom{l-1+p}{p}\sin^{2p}(\omega/2)\right] \tag{1-63}$$

We investigated an example from this family of filters, with $l=3$ and $\tilde{k}=2$. The coefficients of h and $\tilde{h}$ are presented in table 1.2, and the associated scaling functions and wavelets are plotted figure 1.9.

It is clear, in this first example, that the lengths of the filters h and $\tilde{h}$ are very different. This is typical for all examples in this family of "spline filters".

n	0	± 1	± 2	± 3	± 4
$\left(1/\sqrt{2}\right)h(n)$	45/64	19/64	-1/8	-3/64	3/128
$\left(1/\sqrt{2}\right)\tilde{h}(n)$	1/2	1/4	0	0	0

Table 1.2.: Filter coefficients for spline filters with $l=3$, $k=4$, and $\tilde{k}=2$. **Filters 9-3**.

To make it easier for the reader to distinguish the wavelets used, we will designate the associated filters according to their length. These filters, for example, are 9-3. The first number corresponds to the length of the analysis filter and the second to that of the synthesis filter.

1.6.5.2. A spline-variant with filters of less dissimilar lengths

As in the previous case, we choose $R\equiv 0$ in (1-62) but this time, we factorize the right-hand side of (1-62), breaking up the polynomial of degree $l-1$ in $\sin(\omega/2)$ into a product of two polynomials in $\sin(\omega/2)$ with real coefficients. One polynomial is assigned to $H(\omega)$ and the other to $\tilde{H}(\omega)$, in order that the lengths of h and $\tilde{h}$ are as close as possible.

The example described here yields the smallest possible lengths for filters h and $\tilde{h}$ in this family. It corresponds to $l=4$ and $k=4$. The filter coefficients are given in table 1.3, and the associated scaling functions and wavelets are plotted figure 1.10.

n	0	± 1	± 2	± 3	± 4
$\left(1/\sqrt{2}\right)h(n)$	0,602949	0,266864	-0,078223	-0,016864	0,026749
$\left(1/\sqrt{2}\right)\tilde{h}(n)$	0,557543	0,295636	-0,028772	-0,045636	0

Table 1.3.: Filter coefficients for spline filters (variant with less dissimilar lengths), with $l=3$, $k=4$, and $\tilde{k}=4$. **Filters 9-7**.

It is pointed out that the entries in table 1.3 are truncated decimal expansions of irrational numbers. This differs from examples 1 and 3 in paragraphs 1.6.5.1 and 1.6.5.3., where the $\left(1/\sqrt{2}\right)h(n)$ and are rational.

The scaling function ϕ in this example appears very similar to that presented previously in paragraph 1.6.5.1. However, a closer examination shows that the scaling function in this second example is more regular. In both examples, the wavelets $\tilde{\psi}$ have 4 vanishing moments.

These filters are designated 9-7.

1.6.5.3. Quasi-orthonormal filters

Lastly, there are several examples for which $R \not\equiv 0$. In particular, R can be chosen appropriately such that filters h and $\tilde{h}$ are close in length, and hence close to an orthonormal wavelet filter while maintaining the symmetry and compact support properties, etc... of biorthogonal bases.

Since the Laplacian pyramid is quite popular [15], we chose to build a dual wavelet basis using the Laplacian pyramid filter as the filter $H(\omega)$ or $\tilde{H}(\omega)$. This filter is given by the relationship

$$H(\omega) = -a\,e^{-2i\omega} + 0,25\,e^{-i\omega} + (0,5 + 2a) + 0,25\,e^{i\omega} - a\,e^{2i\omega} \tag{1-64}$$

For image processing applications, the choice of a=0.05 is of interest, since it enables minimum entropy and lowest energy in the sub-images achieved by filtering [15].

In collaboration with I. Daubechies, we were able to propose a dual filter to the case a=0.05, corresponding to $R(\omega) = 48\cos(\omega)/175$ with l=2 and k=2 [4], [6]. The filter coefficients are given in table 1.4, and the associated scaling functions and wavelets are plotted figure 1.11. It is important to note that the scaling functions ϕ and $\tilde{\phi}$ are similar, and therefore ψ and $\tilde{\psi}$ are similar as well. In this case, the filter coefficients are rational.

n	0	± 1	± 2	± 3	± 4
$\left(1/\sqrt{2}\right)h(n)$	0,6	0,25	-0,05	0	0
$\left(1/\sqrt{2}\right)\tilde{h}(n)$	17/28	73/280	-3/56	-3/280	0

Table 1.4.: Filter coefficients for quasi-orthogonal filters. These coefficients are rational and the two filters are close in length. Filter h corresponds to the Laplacian pyramid filter proposed in [15]. In this case, l=2, k=2, and k=2. **Filters 5-7.**

The two biorthogonal filters in this example are similar to an orthonormal wavelet filter of length 6 proposed in [25]. This filter is called a "coiflet" and, since it is associated with an orthonormal wavelet basis, it is non symmetric. It can also be noticed that the filters in this third example are shorter than those presented in the two other families. This is due to the fact that k was chosen small. As in the previous case, these filters are designated 5-7.

A final example in this family corresponds to k=4 and l=4. Filters h and $\tilde{h}$ have respectively 9 and 15 coefficients and both are close to a coiflet filter of length 12. These filters' coefficients are given in table 1.5, and the associated scaling functions and wavelets are plotted figure 1.12. These filters are 9-15.

1.6.6. Characteristics of the filters and associated wavelets

In this section, we compare the qualities of the different wavelet bases investigated for image compression applications. Table 1.6 summarizes these properties for biorthogonal bases. We have also

n	0	± 1	± 2	± 3	± 4
$(1/\sqrt{2})h(n)$	0.575	0.28125	-0.05	-0.03125	0.0125
$(1/\sqrt{2})\tilde{h}(n)$	0.575291895604	0.286392513736	-0.052305116758	-0.039723557692	0.015925480769

n	± 5	± 6	± 7
$(1/\sqrt{2})h(n)$	0	0	0
$(1/\sqrt{2})\tilde{h}(n)$	0.003837568681	-0.001266311813	-0.000506524725

Table 1.5.: Filter coefficients in the case l=4, k=4, and $\tilde{k}$=4. **Filters 9-15.**

included the characteristics of Meyer's wavelet as well as those of an orthogonal wavelet proposed by Daubechies for N=2, presented in paragraph 1.4.2.

This table offers a comparison in terms of numerical criteria on the wavelets themselves, rather than in terms of visual criteria, since there is not yet any mathematical criterion enabling measurement of the visual quality of the processed image.

The listings in the table are the numerical values of the criteria defined in paragraph 1.5, which justified our choice of wavelets for encoding. The results show that the filters called 9-7 present the best characteristics. These filters will be used for our experimental results.

Bases	l	k	$\tilde{k}$	v ψ	$\tilde{\psi}$	Vanishing moments ψ	$\tilde{\psi}$	m_2 (10^{-3}) ψ	$\tilde{\psi}$	Linear phase	Length h	$\tilde{h}$
Filters 9-3	3	4	2	-1,83	-2	2	4	398	316	yes	9	3
Filters 9-7	4	4	4	-1,54	-2,3	4	4	326	345	yes	9	7
Filters 5-7	2	2	2	-1,5	-1,5	2	2	289	297	yes	5	7
Filters 9-15	4	4	4	-2,5	-2,5	4	4	336	324	yes	9	15
Meyer	-	-	-	∞		∞		420		yes	∞	∞
Daubechies N=2	-	-	-	-1,5		2		-		no	4	4

Table 1.6. : Quality of the biorthogonal and orthonormal bases used in image compression

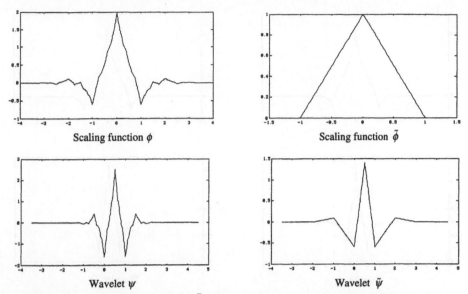

Figure 1.9.: **9-3 filters** - Scaling functions ϕ, $\tilde{\phi}$ and wavelets ψ, $\tilde{\psi}$ associated with spline filters with $l=3$, $k=4$ and $\tilde{k}=2$.

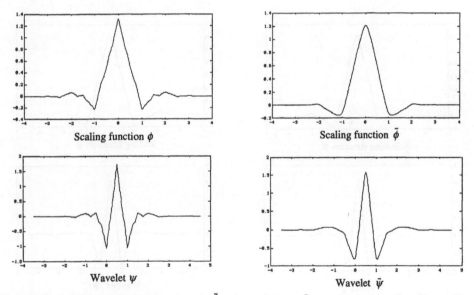

Figure 1.10.: **9-7 filters** - Scaling functions ϕ, $\tilde{\phi}$ and wavelets ψ, $\tilde{\psi}$ associated with spline filters of less dissimilar length, with $l=4$, $k=4$ and $\tilde{k}=4$.

92

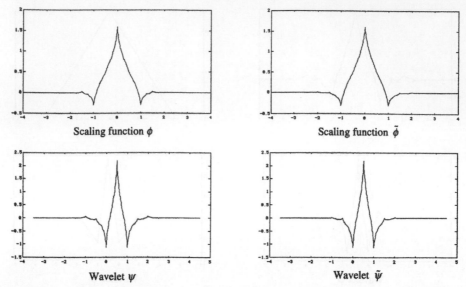

Figure 1.11.: **5-7 filters** - Scaling functions ϕ, $\tilde{\phi}$ and wavelets ψ, $\tilde{\psi}$ associated with quasi-orthogonal filters, with $l=2$, $k=2$ and $\tilde{k}=2$

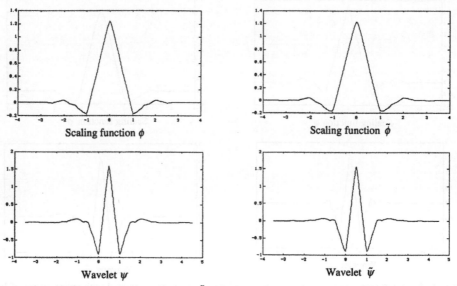

Figure 1.12.: **9-15 filters** - Scaling functions ϕ, $\tilde{\phi}$ and wavelets ψ, $\tilde{\psi}$ associated with filters corresponding to $l=4$, $k=4$ and $\tilde{k}=4$.

1.7. EXTENSION TO THE BIDIMENSIONAL CASE

Wavelet theory can easily be generalized to any dimension $n \in \mathbb{N}^*$ [44]. Here, we examine the special case $n=2$ for image processing applications. As in the case of one-dimensional signals, a set of subspaces is defined $(V_m)_{m \in \mathbb{Z}}$ in $\mathsf{L}^2(\mathbb{R}^2)$ and it can be shown that a scaling function $\phi(x,y)$ exists such that the family of functions $(\phi_{m,n}(x,y))_{n \in \mathbb{Z}^2}$ forms a basis of V_m. The approximation of a signal $f(x,y) \in \mathsf{L}^2(\mathbb{R}^2)$ is thus given by its projection onto the corresponding subspaces V_m.

There are many methods for extending the wavelet transform from the one-dimensional to the two-dimensional case. In the present, we investigated and utilized two particular cases of two-dimensional multiresolution analysis. In paragraph 1.7.1., we describe the separable dyadic multiresolution analysis introduced by Mallat [37], and in paragraph 1.7.2. multiresolution analysis with a resolution factor $\sqrt{2}$, developed in [28], [29] in a similar manner to the quincunx transform [2].

1.7.1. Separable dyadic multiresolution analysis

In this case, each subspace V_m corresponds to a tensor product of two identical subspaces in $\mathsf{L}^2(\mathbb{R})$

$$\mathcal{V}_m(x,y) = V_m(x) \otimes V_m(y) \tag{1-65}$$

It can be shown that the sequence of spaces $(\mathcal{V}_m)_{m \in \mathbb{Z}}$ forms a multiresolution analysis in $\mathsf{L}^2(\mathbb{R}^2)$ if and only if $(V_m)_{m \in \mathbb{Z}}$ is a multiresolution analysis in $\mathsf{L}^2(\mathbb{R})$. Thus the scaling function $\phi(x,y)$ is expressed as the product of two one-dimensional scaling functions

$$\phi(x,y) = \phi(x)\,\phi(y) \tag{1-66}$$

where $\phi(x)$ is the one-dimensional scaling function belonging to $(V_m)_{m \in \mathbb{Z}}$. The approximation of a signal $f(x,y)$ at resolution 2^{-m} is then given by the following relation:

$$s_m(n_x, n_y) = \left(\left\langle f(x,y), \phi_{m,n_x}(x)\,\phi_{m,n_y}(y) \right\rangle \right)_{(n_x, n_y) \in \mathbb{Z}^2} \tag{1-67}$$

As in the one-dimensional case, the detail signal is achieved by projecting $f(x,y)$ onto the complementary subspace defined by $(W_m)_{m \in \mathbb{Z}}$. A basis in W_m can be constructed by translating and dilating three wavelet functions [37]. Let $\psi(x)$ be the wavelet associated with $\phi(x)$, then the three wavelets

$$\begin{aligned}
\psi^H(x,y) &= \phi(x)\,\psi(y) \\
\psi^V(x,y) &= \psi(x)\,\phi(y) \\
\psi^D(x,y) &= \psi(x)\,\psi(y)
\end{aligned} \tag{1-68}$$

are such that $\left(\psi_{m,n}^d\right)_{n\in\mathbb{Z}^2}$ (d=contours orientation) is a basis in $\mathcal{W}_m^d$ with

$$\mathcal{W}_m = \mathcal{W}_m^H \oplus \mathcal{W}_m^V \oplus \mathcal{W}_m^D$$

complements which may or may not be orthogonal, depending on whether the multiresolution analysis is orthogonal or biorthogonal.

The difference between two consecutive approximations of $f(x,y)$ is therefore characterized by three detail signals

$$c_m^H\left(n_x,n_y\right) = \left(\left\langle f(x,y), \phi_{m,n_x}(x)\,\psi_{m,n_y}(y)\right\rangle\right)_{(n_x,n_y)\in\mathbb{Z}^2}$$

$$c_m^V\left(n_x,n_y\right) = \left(\left\langle f(x,y), \psi_{m,n_x}(x)\,\phi_{m,n_y}(y)\right\rangle\right)_{(n_x,n_y)\in\mathbb{Z}^2} \qquad (1\text{-}69)$$

$$c_m^D\left(n_x,n_y\right) = \left(\left\langle f(x,y), \psi_{m,n_x}(x)\,\psi_{m,n_y}(y)\right\rangle\right)_{(n_x,n_y)\in\mathbb{Z}^2}$$

In practice, the images to be processed are sampled signals; the original image, at resolution 2^0, is called $s_0\left(n_x,n_y\right)$. Computation of the image $s_m\left(n_x,n_y\right)$ at a lower resolution 2^{-m} and determination of the wavelet coefficients $c_m^d\left(n_x,n_y\right)$ which are interpreted as three inter-scale images is achieved by convolution operations using separable two-dimensional filters (see figure 1.14.). Filtering is applied independently on the image rows and columns based on the algorithm defined in paragraphs 1.4.1.3. and 1.6.3. for one-dimensional signals.

For a given resolution level M, the original image is represented as a set of $3M+1$ subimages (see figure 1.13):

$$\{s_M\} \cup \left\{\left(c_m^H\right)_{M\le m\le 1}\right\} \cup \left\{\left(c_m^V\right)_{M\le m\le 1}\right\} \cup \left\{\left(c_m^D\right)_{M\le m\le 1}\right\}$$

and this set of subimages is the dyadic two-dimensional wavelet representation of the original image. Because of subsampling, the total number of pixels of this set of subimages is, of course, the same as in the initial image. This separable multiresolution analysis method enables us to distinguish the Horizontal $\left(c_m^H\right)$, Vertical $\left(c_m^V\right)$, and Diagonal $\left(c_m^D\right)$ edges. Consecutive scales are separated by a resolution factor of 2, which explains why the resolution is called "dyadic".

It is to be noted that the actual implementation of the filtering scheme depicted in figure 1.14. involves using analysis filters which are normalized to 1 and synthesis filters with a gain of 2. Indeed, for reasons related to the storage and display of processed images, these filters are not normalized to $\sqrt{2}$.

As an example, we present, in figure 1.15. separable two-dimensional scaling functions $\phi(x,y)$ and $\tilde{\phi}(x,y)$ Figure 1.23. shows the dyadic multiresolution decomposition of the Lena image into two resolution levels, as in the scheme depicted in figure 1.13. The wavelet coefficient subimages are normalized over the interval [0, 255] to be visualized.

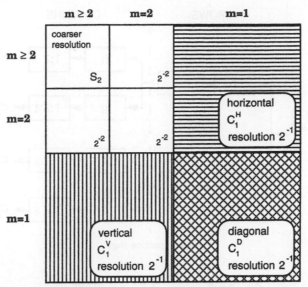

Figure 1.13.: Dyadic wavelet representation of an image: organization of the detail subimages and the subimage at lower resolution. This example corresponds to a decomposition from scale 2^0 to scale 2^2.

1.7.2. Multiresolution analysis with a resolution factor of $\sqrt{2}$

1.7.2.1. Principle

This concept was introduced by Feauveau [28] based on work conducted by Vetterli [52] and Adelson and Simoncelli [2] on the quincunx pyramidal transform. This paragraph presents an overview of the theory and the quincunx multiresolution algorithm used in digital image processing. This algorithm uses non-separable and non-oriented filters. It enables the original image to be decomposed with a resolution factor of $\sqrt{2}$, which means that this analysis is twice as fine as that presented in paragraph 1.8.1. in the two-dimensional case. Furthermore, a single interscale image of wavelet coefficients is sufficient whereas three were required in the dyadic case. Here, a biorthogonal version of this wavelet transform is utilized. For more extensive treatment of this topic, the reader is referred to [28], [29]

The two-dimensional scaling function which gives rise to multiresolution analysis with a resolution factor of $\sqrt{2}$, is defined, $\forall m \in (1/2)\mathbf{Z} = \{i/2; i \in \mathbf{Z}\}$, by

$$\phi_{m,n}(x,y) = 2^{-m}\phi\left(L^{-2m}(x,y)-n\right) \qquad n=\left(n_x,n_y\right)\in \mathbf{Z}^2 \qquad (1\text{-}70)$$

where $L(x,y)=(x+y,x-y)$ is a linear transform such that $L^n = L o L ... L$, o designating the law of composition of applications. This transform preserves angles and dilates distances by $\sqrt{2}$.
In addition, a filter function $H(\omega_x,\omega_y)$ in $\mathsf{L}^2([0,2\pi[)$ is introduced such that

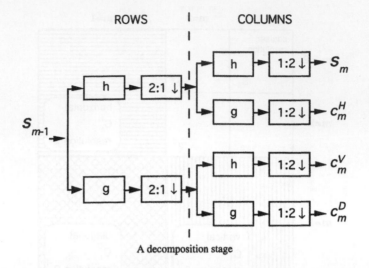

A decomposition stage

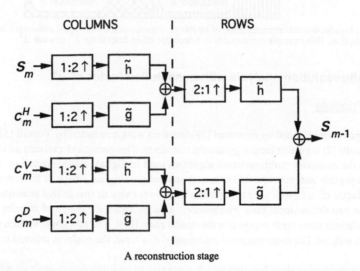

A reconstruction stage

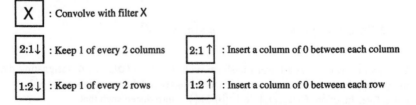

Figure 1.14.: Multiscale decomposition/reconstruction of an image.

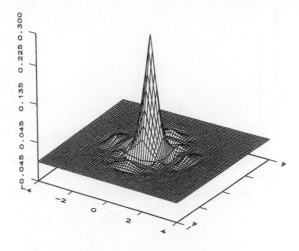

Scaling function $\phi(x,y)=\phi(x)\phi(y)$

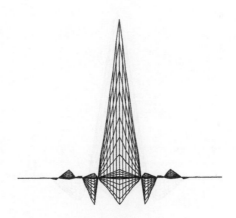

Left view of the scaling function $\phi(x,y)$

Figure 1.15.a.: Separable two-dimensional scaling function $\phi(x,y)$ constructed from the one-dimensional 9-7 filters described in paragraph 1.6.5.2.

98

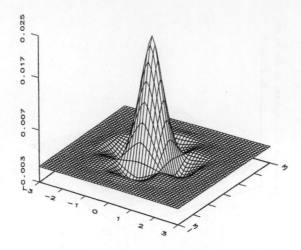

Scaling function $\tilde{\phi}(x,y) = \tilde{\phi}(x)\tilde{\phi}(y)$

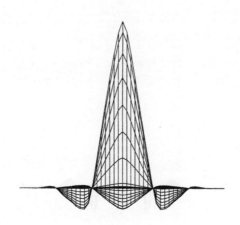

Left view of the scaling function $\tilde{\phi}(x,y)$

Figure 1.15.b.: Separable two-dimensional scaling function $\tilde{\phi}(x,y)$ constructed from the one-dimensional 9-7 filters described in paragraph 1.6.5.2.

$$\Phi\big(\omega_x + \omega_y, \omega_x - \omega_y\big) = H\big(\omega_x, \omega_y\big)\ \Phi\big(\omega_x, \omega_y\big) \tag{1-71}$$

The wavelet $\psi(x,y)$ is then defined by its Fourier transform [28]

$$\Psi\big(\omega_x + \omega_y, \omega_x - \omega_y\big) = \tilde{H}\big(\omega_x + \pi, \omega_y + \pi\big)\ e^{-i\omega_x}\ \Phi\big(\omega_x, \omega_y\big) \tag{1-72}$$

where $\tilde{H}$ is the dual of H defined in paragraph 1.6.

Thus, the interscale information can be extracted by a single wavelet rather than three wavelets in the dyadic two-dimensional case.

1.7.2.2. Construction of two-dimensional filters - Filtering

We computed our filters $h(x,y)$ and $\tilde{h}(x,y)$ from the one-dimensional filters defined in paragraph 1.6.5. These filters are extended to the two-dimensional case by applying a Mac Clellan type transform

$$\cos(\omega) \rightarrow \frac{1}{2}\big(\cos(\omega_x) + \cos(\omega_y)\big) \tag{1-73}$$

such that equation (1-62) is verified. This transform ensures that all of the properties of one-dimensional filters are also exhibited by two-dimensional filters.

As an example, table 1.7. presents the coefficients of the 9-7 filters described in paragraph 1.6.5.2, extended to the case of two-dimensional multiresolution analysis with a resolution factor of $\sqrt{2}$. The diagram to the left of the table shows the arrangement of the non-null coefficients of a square two-dimensional filter. The other filter coefficients have values equal to zero and are therefore not depicted.

The filtering algorithm using this type of filter is given schematically in figure 1.18. This algorithm is described in detail in [28], [29].

The non-separable two-dimensional scaling functions $\phi(x,y)$ and $\tilde{\phi}(x,y)$, are shown in figure 1.19.

Lastly, the organization of the subimages resulting from the analysis is given in figure 1.17. for a decomposition up to resolution 2^{-2}. To visualize the transform, and for efficient use of the image coding transform, the coefficients $c_{h-1/2}$ $h \in Z$ are rearranged in a more compact manner [28]. This involves reorganizing the coefficients which, after filtering, are on a spatial grid in quincunx onto a Z^2 grid to facilitate visualization.

Figure 1.24 shows the Lena quincunx wavelet decomposition, and figure 1.25 presents the decomposition of the same image according to the scheme in figure 1.17.

1.7.3. Boundering effects

Digital filtering always gives rise to boundering effects, due to the spatial length of the filters. Indeed, the convolutions performed in this case are somewhat delicate, since all of the edge information in the processed image must ideally be preserved.

100

Consider, for example, an image row of N columns; the pixel value is written x_i where i is the column index :

$$x_1, x_2, x_3, \ldots, x_{N-2}, x_{N-1}, x_N$$

There is generally a discontinuity when, considering periodicity along the row, x_N is followed by $x_1, x_2, x_3, \ldots$. During filtering, this discontinuity introduces parasitic frequencies which perturb the reconstruction algorithm [12], [14] .

Since we are using filters exhibiting even symmetry, one solution which has already been proposed by our laboratory [12], [13], [9] consists in making the image row symmetrical as follows:

$$x_1, x_2, x_3, \ldots, x_{N-2}, x_{N-1}, x_N, z, x_N, x_{N-1}, x_{N-2}, \ldots, x_2, x_1$$

This new row is obtained by copying the pixel values from x_2 to x_N. The value of the additional point z is interpolated by a polynomial, for example. Thus the convolution operation on point z, which is at the edge of the image, can now be performed without introducing any visual artefacts. This preprocessing procedure is performed on all of the rows and columns in the original image. The interpolated points z are filtered and kept throughout the decomposition, in order to avoid significant propagation of interpolation errors.

Figure 1.16. shows the symmetry created around the image before dyadic multiresolution analysis, or with a resolution factor $\sqrt{2}$. In the dyadic case, the corners of the image are not considered.

Interpolation
z

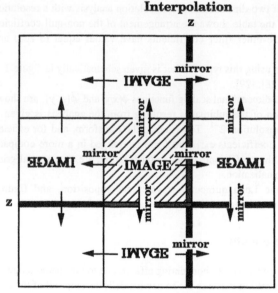

Figure 1.16.: Symmetry created around the image.

		h	$\tilde{h}$
a	a	0.001671	—
f b f	b	-0.002108	-0.005704
j g c g j	c	-0.019555	-0.007192
f g i d i g f	d	0.139756	0.164931
a b c d e d c b a	e	0.687859	0.586315
f g i d i g f	f	0.006687	—
j g c g j	g	-0.006324	-0.017113
f b f	i	-0.052486	-0.014385
a	j	0.010030	—

Table 1.7.: Two-dimensional filter coefficients

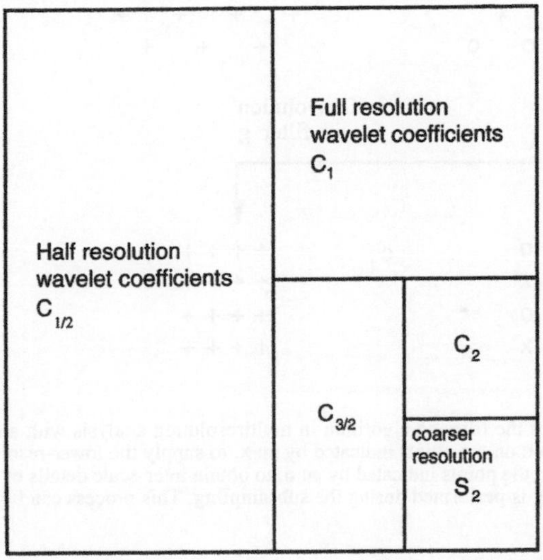

Figure 1.17.: Organization of detail subimages and the coaser resolution subimage (at the immediately lower resolution) in a non-separable two-dimensional wavelet representation. This example corresponds to a decomposition from scale 2^0 to scale 2^2.

102

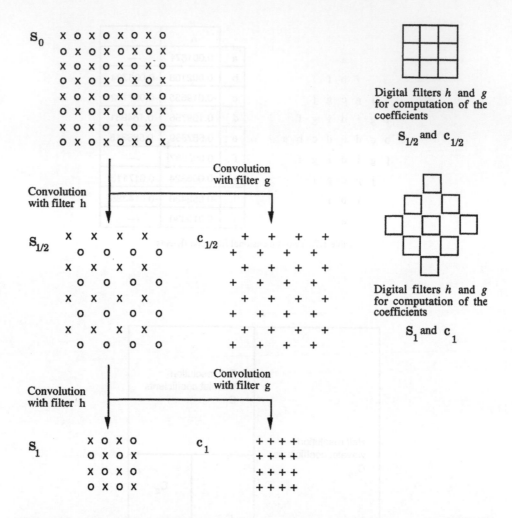

Figure 1.18.: Flowchart of the filtering algorithm in multiresolution analysis with a resolution factor $\sqrt{2}$. The low-pass filter h is centered on the points indicated by an X, to supply the lower-resolution image, and the high-pass filter g is centered on the points indicated by an o, to obtain inter-scale details or wavelet coefficients. Thus the shift caused by filter g is performed during the subsampling. This process can be iterated again from image S_1.

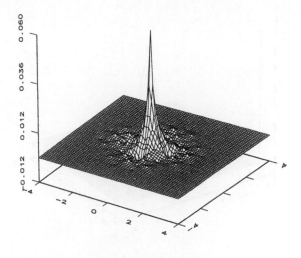

Scaling function $\phi(x, y)$

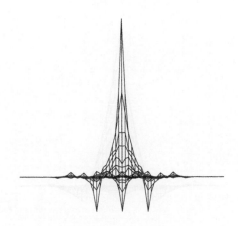

Left view of the scaling function $\phi(x, y)$

Figure 1.19.a.: Non-separable two-dimensional scaling function $\phi(x, y)$ corresponding to the filters presented in Table 1.7.

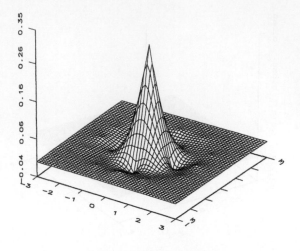

Scaling function $\tilde{\phi}(x,y)$

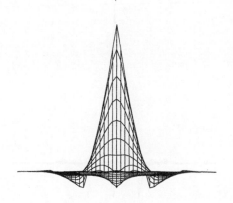

Left view of the scaling function $\tilde{\phi}(x,y)$

Figure 1.19.b.: Non-separable two-dimensional scaling function $\tilde{\phi}(x,y)$, corresponding to the filters presented in Table 1.7.

Note that an other solution was introduced by Cohen, Daubechies and Vial [19] to avoid the boundering effect. In order to adapt the wavelet transform to images, they introduced wavelets with compact support on an interval. In that scheme, we must define new filters for the right and left bounds of the image. The number of these filters increases according the number of wanishing moments of the "classical" wavelet, i.e., the wavelet used on the whole image. This results in an increasing of the complexity of the wavelet transform, which is not interesting in image coding applications [34].

1.8. STATISTICAL PROPERTIES OF WAVELET COEFFICIENTS

The type of quantizer to be used as well as its image coding performances can be determined from a statistical analysis of the subimages obtained after the wavelet transform. The normalized histogram, or probability density function (pdf) of a subimage provides us with information on the distribution of the coefficient values in this subimage (mean, standard deviation...). It is therefore important to model this histogram in order to gain a quantitative description of the quantifier performances. Much work in this field has been described in the literature [44], [30] based on various approximation models.

In figure 1.20, we present a typical pdf of a wavelet coefficient subimage. This pdf was plotted from a dyadic wavelet transform applied to the Lena image with 9-7 filters. It corresponds to resolution 2^{-1} and to vertical edges. Similar pdfs can be obtained for different types of images with different wavelets. The use of a wavelet transform with a resolution factor $\sqrt{2}$, for example, yields similar results.

With respect to the pdf of the original image, which is generally multimode and difficult to analyse, this pdf is single-mode and quite narrow. Its variance is low and its mean is null, which indicates that in each subimage, there are many wavelet coefficients with low values.

The pdfs of wavelet coefficient subimages, which are pointed and narrow, immediately bring into mind Laplace laws [42]. However, these laws do not offer a satisfactory model for this type of pdf, particularly around the origin. For this reason, the mathematical model selected is the *generalized Gaussian* [4], [6]. This approximation law enables us to refine the modeling of the decay around the origin as well as the pdf's tails, thus making it a more accurate model of the observed pdf.

The generalized Gaussian is given by [1]

$$p_X(x) = a e^{-|bx|^\alpha} \tag{1-74}$$

where $a = \dfrac{b\alpha}{2\Gamma(1/\alpha)}$ and $b = \dfrac{1}{\sigma}\sqrt{\dfrac{\Gamma(3/\alpha)}{\Gamma(1/\alpha)}}$ \hfill (1-75)

where σ is the standard deviation of the pdf to be modeled and the Gamma function given by

$$\Gamma(\beta) = \int\limits_{0}^{+\infty} e^{-x} x^{\beta-1} dx \tag{1-76}$$

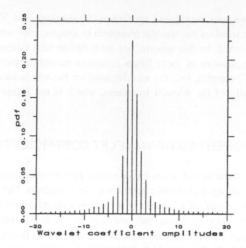

Figure 1.20.: Typical subimage probability density function

Note that the general formula (1-74) encompasses, as particular cases, the Laplace law for $\alpha=1$ and the Gaussian law for $\alpha=2$.

In order to adjust the observed distribution to a theoretical law, we used the following formula

$$t_{\chi^2} = \sum_{i=1}^{K} \frac{\left(p_X(x_i) - p(x_i)\right)^2}{p(x_i)} \tag{1-77}$$

where $p_X(x_i)$ corresponds to the discrete law of probability observed and $p(x_i)$ to the theoretical approximation model. K is the number of amplitude values taken by the wavelet coefficients. Note that this formula is the same as that used in the χ^2 test. However, this test was not used in the present case to examine the model's validity, but simply to adjust the model to the observed pdf.

To approximate the pdfs of a given image with variance σ^2, we seek the value of which minimizes the criterion t_{χ^2}. As an example, we have plotted in figure 1.21. how t_{χ^2} varies with the vertical edges subimage of the Lena image, at resolution 2^{-2}. Here, the value corresponds to the statistical minimum of the criterion.

Figure 1.22. shows the wavelet coefficient subimage pdf plotted at resolution 2^{-3} for three different approximation models: Gaussian, Laplacian, and generalized Gaussian (with $\alpha=0.6$). These three theoretical models are adjusted to the subimage variance. The generalized Gaussian obviously provides the best approximation of the observed pdf, considering that the tails of the observed pdf are long.

The results of tests conducted on the other wavelet coefficient subimages obtained from the Lena image are summarized in table 1.8. These tests were performed on subimages up to resolution 2^{-4}.

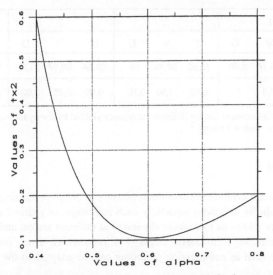

Figure 1.21.: Variation of the criterion t_{χ^2} with α (variation increments of 0.05).

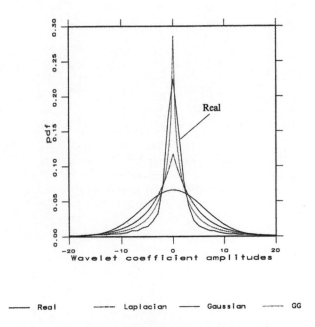

Figure 1.22.: Observed pdf (Real) and three different approximation models: the Generalized Gaussian (GG) with $\alpha=0.6$, the Laplacian ($\alpha=1$) and the Gaussian law ($\alpha=2$).

m	1			2			3			4		
Orientation	H	V	D	H	V	D	H	V	D	H	V	D
σ^2	7,66	14,76	3,66	13,53	36,34	10	20,84	66,97	21,35	28,23	111	27,67
α_{χ^2}	0,85	0,8	1	0,65	0,60	0,70	0,60	0,55	0,60	0,65	0,65	0,60

Table 1.8.: Results of the adjustment test for different subimages yielded by the dyadic wavelet transform applied to the Lena image, with 9-7 filters.

1.9. CONCLUSION

An essential driving factor behind multiresolution analysis in image processing, based on the use of digital filters, is the ability to handle separately each subimage, or spatio-temporal element. The multiresolution concept allows us to consider the image at different scales, and hence to distinguish fine edges (high resolution) from coarse edges (low resolution). Thus, for a given subimage, processing operations such as compression or coding can be adapted to the resolution level and statistical distribution.

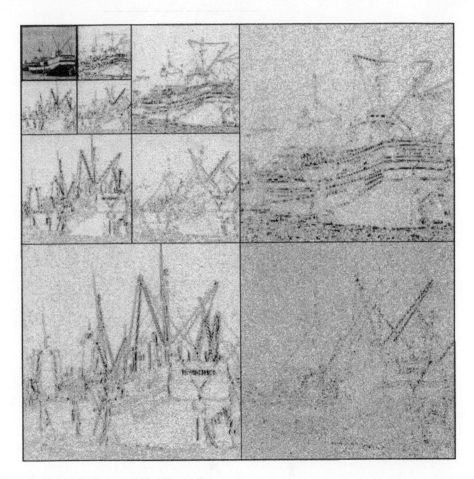

Figure 1.23.: Dyadic decomposition of image Boat.

110

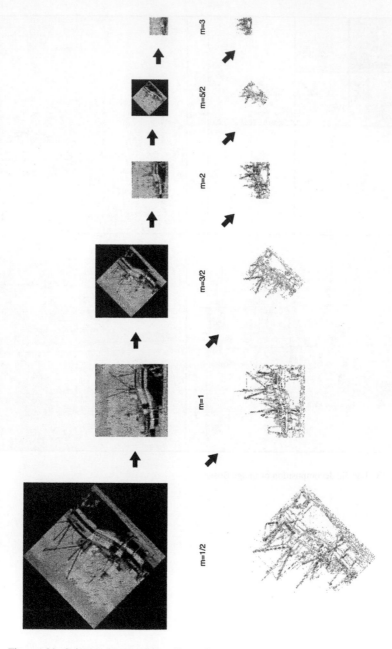

Figure 1.24.: Quincunx decomposition of image Boat

Figure 1.25.: Quincunx decomposition of image Boat - organization of subimages according figure 1.17.

APPENDIX A

A BRIEF REVIEW OF DIGITAL FILTER BANKS

The application of filter banks to sub-band speech coding was first introduced in 1976 by Crochiere et al [20]. At about the same time, Croisier, Esteban, and Galand [21] proposed a family of special filters called Quadrature Mirror Filters (QMF) because their respective phases are in quadrature. The basic principle consists in breaking down a signal into different frequency bands and processing each band separately since the implementation of this type of filter is relatively easy.

A.1. FILTERING BY FILTER BANKS

The basic filter bank unit is a pair of filters, a low-pass filter and a high-pass filter. The input signal is decomposed into two frequency *subbands*, a low frequency subband and a high frequency subband. The general structure of a bank of two one-dimensional filters is shown in figure A.1.

The sampled input signal $x(n)_{n \in Z} \in l^2(Z)$ is filtered respectively by a low-pass filter $h(n)_{n \in Z}$ and a high-pass filter $g(n)_{n \in Z}$. These digital filters are called <u>analysis filters</u>. The resulting signals $x_b(n)$ and $x_h(n)$ each occupy half the bandwidth of the input signal. Using Shannon's theorem, signals $y_b(n)$ and $y_h(n)$ are obtained by subsampling $x_b(n)$ and $x_h(n)$ by a factor 2 (decimation). The subsampling of these signals at their Nyquist frequency does not cause spectral folding since the filters are QMF. Furthermore, with subsampling, the number of samples of signals of $y_b(n)$ and $y_h(n)$ are respectively half the number of the input signal $x(n)$. Note that the total number of samples after filtering into two subbands has not increased, and that it remains the same as that of the input signal.

After processing (coding, transmission, and decoding, for example), the signal is reconstructed from the processed signals $\hat{y}_b(n)$ and $\hat{y}_h(n)$. These signals are oversampled in order to obtain the signals $\hat{x}_b(n)$ and $\hat{x}_h(n)$, which are then filtered by a low-pass filter $\tilde{h}(n)$ and high-pass filter $\tilde{g}(n)$, respectively. These filters are called <u>synthesis filters</u>. Lastly, the two resulting signals are added to obtain the reconstructed signal $\hat{x}(n)$.

The choice of the filter pairs h, $\tilde{h}$ and g, $\tilde{g}$ is critical and must take into account certain constraints, as will be discussed in the following paragraphs.

A.2. QUADRATURE MIRROR FILTERS (QMF)
AND CONJUGATE QUADRATURE FILTERS (CQF)

A.2.1. INPUT/OUPUT Relationships

The signals obtained after filtering by filters $h(n)$ and $g(n)$ are given by:

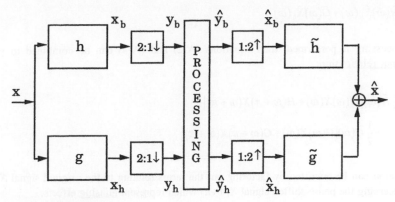

X	: Convolve with filter X
2:1↓	: Keep one sample out of two (undersampling)
1:2↑	: Put a zero between each sample (oversampling)

Figure A.1.: Principle of signal Analysis/Synthesis by filter banks.

$$X_b(\omega) = H(\omega)X(\omega)$$
$$X_h(\omega) = G(\omega)X(\omega) \tag{A-1}$$

where $H(\omega)$ and $G(\omega)$ are the respective Fourier transforms of $h(n)$ and $g(n)$, and $X(\omega)$ is the Fourier transforms of $x(n)$.

After subsampling, the subband signals are expressed as follows:

$$Y_b(\omega) = \frac{1}{2}\left[H\left(\frac{\omega}{2}\right)X\left(\frac{\omega}{2}\right) + H\left(\frac{\omega}{2}+\pi\right)X\left(\frac{\omega}{2}+\pi\right) \right]$$
$$Y_h(\omega) = \frac{1}{2}\left[G\left(\frac{\omega}{2}\right)X\left(\frac{\omega}{2}\right) + G\left(\frac{\omega}{2}+\pi\right)X\left(\frac{\omega}{2}+\pi\right) \right] \tag{A-2}$$

and after processing, the signals interpolated by a factor 2 are given by:

$$\hat{X}_b(\omega) = \hat{Y}_b(2\omega)$$
$$\hat{X}_h(\omega) = \hat{Y}_h(2\omega) \tag{A-3}$$

Finally, the reconstructed signal is written:

$$\hat{X}(\omega) = \tilde{H}(\omega)\hat{X}_b(\omega) + \tilde{G}(\omega)\hat{X}_h(\omega) \qquad (A-4)$$

If no processing is performed, then equations (A-2) and (A-4) can be combined to yield the input/output relationship:

$$\hat{X}(\omega) = \frac{1}{2}\tilde{H}(\omega)\big[H(\omega)X(\omega) + H(\omega + \pi)X(\omega + \pi)\big]$$
$$+ \frac{1}{2}\tilde{G}(\omega)\big[G(\omega)X(\omega) + G(\omega + \pi)X(\omega + \pi)\big] \qquad (A-5)$$

This equation can be rewritten by factoring out the terms relating to the original signal $X(\omega)$ and those concerning the phase-shifted signal $X(\omega + \pi)$ which represent aliasing effects:

$$\hat{X}(\omega) = \frac{1}{2}\big[\tilde{H}(\omega)H(\omega) + \tilde{G}(\omega)G(\omega)\big]X(\omega)$$
$$+ \frac{1}{2}\big[\tilde{H}(\omega)H(\omega + \pi) + \tilde{G}(\omega)G(\omega + \pi)\big]X(\omega + \pi) \qquad (A-6)$$

To ensure exact reconstruction, it is clear that the second term in equation (A-6) (i.e., the factor of $X(\omega + \pi)$) must be equal to 0, in order to eliminate spectral folding. In addition, the first term, the factor of $X(\omega)$ must be equal to 2 so that $X(\omega) = \hat{X}(\omega)$. The exact reconstruction equation is thus written:

$$\tilde{H}(\omega)H(\omega) + \tilde{G}(\omega)G(\omega) = 2 \qquad (A-7)$$

and the aliasing cancellation equation

$$\tilde{H}(\omega)H(\omega + \pi) + \tilde{G}(\omega)G(\omega + \pi) = 0 \qquad (A-8)$$

Esteban and Galand [27], and later Smith and Barnwell [48], proposed relationships between filters H and $\tilde{H}$ and G and $\tilde{G}$, based on equation (A-7). These relationships are presented briefly hereafter. A more detailed description of QMFs and CQFs can be found in [27], [48], and [58].

A.2.2. Quadrature Mirror Filters (QMF)

Quadrature Mirror Filters are used for speech and image coding. To eliminate spectral folding during signal synthesis while satisfying equation (A-8) (aliasing cancellation), the following relationships must be verified:

$$\begin{cases} G(\omega) = H(\omega + \pi) \\ \tilde{H}(\omega) = G(\omega + \pi) \\ \tilde{G}(\omega) = -H(\omega + \pi) \end{cases} \qquad (A\text{-}9)$$

The QMF term results from the fact that the high-pass filter $G(\omega)$ response is the mirror image of that of the low-pass filter $H(\omega)$.

The exact reconstruction relationship is then

$$H^2(\omega) - H^2(\omega + \pi) = 2 \qquad (A\text{-}10)$$

However, it is pointed out that it is impossible to obtain QMFs which are exact reconstruction FIR (Finite Impulse Response) filters.

A.2.3. Conjugate Quadrature Filters (CQF)

Smith and Barnwell [48] proposed a different type of filter from that presented in formula (A-9) which also enables exact reconstruction by eliminating spectral folding. The relationships for CQFs are given below:

$$\begin{cases} G(\omega) = -e^{-in\omega} \overline{H}(\omega + \pi) \\ \tilde{H}(\omega) = G(\omega + \pi) \\ \tilde{G}(\omega) = -H(\omega + \pi) \end{cases} \qquad (A\text{-}11)$$

The exact reconstruction relationship is

$$|H(\omega)|^2 + |H(\omega + \pi)|^2 = 2 \qquad (A\text{-}12)$$

The advantage of CQFs over QMFs is that the reconstruction is rigorously exact. However, CQFs built using FIR filters are not symmetrical and their implementation is more complex than that of QMFs.

BIBLIOGRAPHY

[1]   M. Abramowitz, I.A. Stegun, "Handbook of Mathematical Functions", *Dover publications, Inc., New-York*, 1965

[2]   E.H. Adelson, E. Simoncelli, R. Hingorani, "Orthogonal Pyramid Transforms for Image Coding", *SPIE Visual Communication and Image Processing II*, Vol. 845, pp. 50-58, 1987.

[3]   N. Ahmed, T. Natarajan and K.R. Rao, "Discrete Cosine Transform", *IEEE Trans. Comput.*, Vol. C-23, pp. 90-93, 1974.

[4]   M. Antonini, M. Barlaud, P. Mathieu, and I. Daubechies, "Image Coding Using Vector Quantization in the Wavelet Transform Domain", *IEEE ICASSP, Albuquerque* USA, pp. 2297-2300, April 1990.

[5]   M. Antonini, "Transformée en Ondelettes et Compression Numérique des Images", *ph-D Dissertation*, University of Nice-Sophia Antipolis, FRANCE, September 1991.

[6]   M. Antonini, M. Barlaud, P. Mathieu, I. Daubechies, "Image Coding Using Wavelet Transform", *IEEE Trans. on Image Processing*, Vol.1, No.2, 1992.

[7]   M. Antonini, M. Barlaud, P. Mathieu, and J.C. Feauveau, "Multiscale Image Coding using the Kohonen Neural Network", *SPIE Visual Communication and Image Processing, Lausanne*, SUISSE, 1990.

[8]   N. Baaziz, C. Labit, "Transformations pyramidales d'Images Numériques", IRISA, *Technical report* n° 526, Mars 1990.

[9]   M. Barlaud, L. Blanc-Féraud, P. Mathieu, J. Menez, M. Antonini, "2D Linear Predictive Image Coding With Vector Quantization", *EUSIPCO, Grenoble*, FRANCE, pp. 1637-1640, Sept. 5-8, 1988.

[10]   M. Barlaud, T. Gaidon, P. Mathieu, and J.C. Feauveau, "Edge Detection using Recursive Biorthogonal Wavelet Transform", *IEEE ICASSP, Toronto, Ontario*, CANADA, May 14-17, 1991.

[11]   G. Battle, "A Block Spin Construction of Wavelets. Part I Lemarié Functions", *Comm. Math. Phys.* 110 (1987) pp. 601-615.

[12]   L. Blanc-Féraud, M. Barlaud, P. Mathieu, "Amélioration de la Restauration d'Images Floues par un Filtrage de Kalman Utilisant une Image Miroir", *Traitement du Signal*, vol.5, pp.249-261, 1988.

[13]   L. Blanc-Féraud, M. Barlaud, P. Mathieu "Ringing Reduction in Images Restoration Using Mirror Images and Adaptive Kalman Filtering", *IEEE ICASSP, New York*, USA, pp. 1012-1015, 1988.

[14] L. Blanc-Féraud, "Modélisation d'image. Application à la Compression Numérique d'Images Floues", *ph-D Dissertation*, University of Nice-Sophia Antipolis, FRANCE, July 1989.

[15] P. Burt and E. Adelson, "The Laplacian Pyramid as a Compact Image Code", *IEEE Trans. Comm.* Vol.31 (1983) 482-540.

[16] G. Beylkin, R. Coifman and V. Rokhlin, "Fast wavelet transforms and numerical analysis, I", *Comm. on Pure and Applied Math.*, 44, pp. 141-183, 1991.

[17] A. Cohen, "Ondelettes, Analyse Multirésolution et Filtres Miroirs en Quadrature", *Institut Henri Poincaré*, non linear analysis, 1990.

[18] A. Cohen, I. Daubechies and J.C. Feauveau, "Biorthogonal Bases of Compactly Supported Wavelets", *AT&T Bell Laboratories, Technical report*, n° TM 11217-900529-07.

[19] A. Cohen, I. Daubechies and P. Vial, "Wavelets and Fast Wavelet Transforms on the Interval", *preprint AT&T Bell Laboratories*, 1992.

[20] R.E. Crochiere, S.A. Webber and J.L. Flanagan, "Digital Coding of Speech in Sub-bands", *Bell Systems Technical Journal*, Vol. 55, No.8, pp.1069-1085, October 1976.

[21] A. Croisier, D. Esteban and C. Galand, "Perfect Channel Splitting by use of Interpolation/Decimation/Tree Decomposition Techniques", *Proc. of the Intern. Conf. on Information Science and Systems*, Patras, Greece, pp.443-446, August 1976.

[22] I. Daubechies, A. Grossmann and Y. Meyer, "Painless Nonorthogonal Expansions", *Journal of Math. Phys. 27* (1986) 1271-1283.

[23] I. Daubechies, "Ten Lectures on Wavelets", *CBMS-NSF Regional Conferences Series in Applied Mathematics 61, SIAM Press*, Philadelphia, 1992.

[24] I. Daubechies, "Orthonormal Bases of Compactly Supported Wavelets", *Comm. Pure Appl. Math. 41* (1988) 909-996.

[25] I. Daubechies, "Orthonormal Bases of Compactly Supported Wavelets. II. Variations on a Theme, *AT&T Bell Laboratories, Technical report*, n° TM 11217-891116-17.

[26] T. Ebrahimi, T.R. Reed, M. Kunt, "Video Coding Using a Pyramidal Gabor Expansion", *Visual Communication and Image Processing, SPIE , Lausanne*, SUISSE, Vol. 1360, pp. 489-502, 1990.

[27] D. Esteban, C. Galand, "Applications of Quadrature Mirror Filters to Split Band Voice Coding Systems", *Proc. of the Intern. Conf. on ASSP (ICASSP), Hartford*, USA, pp.191-195, May 1977.

[28] J.C. Feauveau, "Analyse Multirésolution par Ondelettes Non Orthogonales et Bancs de Filtres Numériques", *ph-D Dissertation*, University of Paris Sud, FRANCE, January 1990.

[29] J.C. Feauveau, "Analyse Multirésolution pour les Images avec un Facteur de Résolution $\sqrt{2}$", *Traitement du Signal*, vol. 7, n°2, pp.117-128, 1990.

[30] J. Froment, "Traitement d'Images et Applications de la Transformée en Ondelettes", *ph-D Dissertation*, University of Paris IX Dauphine U.F.R Mathématiques de la décision, FRANCE, January 1990.

[31] A. Grossmann and J. Morlet, "Decomposition of Hardy functions Into Square Integrable Wavelets of Constant Shape", *SIAM J. Math Anal. 15* (1984) 723-736.

[32] S. Jaffard, "Algorithmes de Transformation en Ondelettes", *"Ecole polytechnique" course*.

[33] A.K. Jain, "A Sinusoidal Family of Unitary Transforms", *IEEE Trans. on Pattern Analysis and machine intelligence*, Vol. PAMI-1, 1979.

[34] D. Lebedeff, M. Barlaud, P. Mathieu, "Ondelettes à Support Compact sur un Intervalle - Mise en Oeuvre en Traitement d'Images", *Technical report n° 92-41*, I3S/CNRS laboratory, Sophia Antipolis (FRANCE), July 1992.

[35] P.G. Lemarié, "Une Nouvelle Base d'Ondelettes de $L^2(_\shortparallel)$", *Pure and applied mathematics*, Vol.67, pp. 227-238, 1988.

[36] P.G. Lemarié and Y. Meyer, "Ondelettes et Bases Hilbertiennes", *Rev. Mat. Iberoamericana*, Vol. 2, pp. 1-18, 1986.

[37] S. Mallat, "A Theory for Multiresolution Signal Decomposition: The Wavelet Representation", *IEEE Trans on Pattern Anal. and Mach. intel.*, Vol. 11 No.7, July 89.

[38] S. Mallat, "Multiresolution Approximation and Wavelet Orthonormal Bases of $L^2(_\shortparallel)$", *Trans. of the American Mathematical Society*, Vol. 315, pp.69-87, Sept. 1989.

[39] S. Mallat, "Multiresolution Représentations and Wavelets", *Ph. D Dissertation*, GRASP Lab, Dept. of Computer and Information Science, University of Pennsylvania, 1988.

[40] D. Marr, "Vision" *W.H. Freeman and Company, New York*, 1982.

[41] P. Mathieu, M. Barlaud, M. Antonini, "Compression d'Images par Transformée en Ondelette", *12ième colloque GRETSI, Juan les Pins*, FRANCE, June 12-16, 1989.

[42] P. Mathieu, M. Barlaud, M. Antonini, "Compression d'Image par Transformée en Ondelette et Quantification Vectorielle", *Traitement du Signal*, vol. 7, n°2, pp.101-115, 1990.

[43] Y. Meyer, "Principe d'Incertitude, Bases Hilbertiennes et Algèbres d'Opérateurs", *Bourbaki conference*, 1985-1986, nr 662.

[44] Y. Meyer, "Ondelettes et Opérateurs", Tomes I, II et III, *Hermann*, 1990. [Hermann, éditeurs, 293 rue Lecourbe, 75015 Paris].

[45] Y. Meyer, "Ondelettes, Filtres Miroirs en Quadrature et Traitement Numérique de l'Image", University of Madrid conference, february 22, 1989.

[46] O. Rioul, "Ondelettes Régulières : Application à la Compression d'Images Fixes", *ph-D Dissertation*, Ecole Nationale Supérieure des Télécommunications, FRANCE, March 1993.

[47] O. Rioul, "Simple Regularity Criteria for Subdivision Scheme", *SIAM J. Math Anal.*, Vol.23, No.6, pp. 1544-1576, November 1992.

[48] M.J. Smith and D.P. Barnwell, "Exact Reconstruction for Tree-Structured Subband Coders", *IEEE Trans. ASSP 34* (1986) 434-441.

[49] P. Strobach, H. Hölzimmer, M. Buchner, "A Design Technique for Generalized Zonal Masks for DCT-Coefficients", *Digital Signal Processing*, pp. 448-452, North-Holland, 1987.

[50] J.O. Stromberg, "A Modified Haar System and Higher Order Spline Systems", *Conf. in harmonic analysis in honor of Antoni Zygmund*, Vol. II, 475-493; eds W. Beckner et al., Wadworth Math. Series.

[51] P. Tchamitchian, "Biorthogonalité et Théorie des Opérateurs", *Revista. Matematica. Iberoamericana*, Vol. 3, No.2, 1987.

[52] M. Vetterli, "Multidimensional Subband Coding : Some Theory and Algorithms", *Signal Processing* 6, 1984.

[53] M. Vetterli, "Splitting a Signal Into Subsampled Channels Allowing Perfect Reconstruction", *Proc. IASTED Conf. Appl. Signal Processing Digital Filtering,* Paris France, June 1985.

[54] M. Vetterli, "Filter Banks Allowing Perfect Reconstruction", *Signal Processing*, Vol. 10, pp.219-244, 1986.

[55] M. Vetterli, C. Herley, "Wavelets and Filter Banks : Relationships and New Results " *IEEE ICASSP, Albuquerque*, USA, pp. 1723-1726, April 1990.

[56] J. Kovacevic, M. Vetterli, "Design of Multidimensional Non-Separable Regular Filter Banks and Wavelets", *IEEE ICASSP, San Francisco, California*, USA, pp. IV389-IV392, March 23-26,1992.

[57] B.V.K. Vijaya Kumar, C.P. Neuman, K.J. De Vos, "Discrete Wigner Synthesis", Signal Processing, Vol. 11, pp. 277-304, 1987.

[58] P.H. Westerink, D.E. Boekee, J. Biemond, J.W. Woods, "Subband Coding of Image Using Vector Quantization", *IEEE Trans. on Comm.*, Vol. 36, pp. 713-719, 1988.

[59] T. Kronander, "Some Aspect of Perception Based Image Coding", *ph-D Dissertation*, Linköping University, SWEDEN, 1989.

[60] S. Pittner, J. Schneid, C.W. Ueberhuber, "Wavelet Literature Survey", *Institute for Applied and Numerical Mathematics, Technical University Vienna* [Wiedner Hauptstrasse 8-10/115, A-1040 Wien Austria].

[61] R. Ansari, C. Guillemot, "Exact Reconstruction Filter Banks using Diamond FIR Filters", *Proceedings of 1990 Bilkent Conference on New Trends in Communications and Signal Proc*, Ed. E. Arikan, Elsevier, 1990.

[62] M.Kunt, "Traitement Numérique des Signaux", *Dunod*, 1981.

2. VECTOR QUANTIZATION

2.1. INTRODUCTION

Vector quantization is a generalization of scalar quantization to vectors of dimension n. Each vector is comprised of points (coordinates) belonging to the set of real numbers $\mathbf{R}$. Vector quantization consists in quantizing all of the coordinates of a vector instead of separately quantizing each of its coordinates. The extension from one to several dimensions calls upon new concepts and techniques opening the way to applications previously inconceivable in one dimension (exploiting spatial correlation, for example).

While scalar quantization is generally used in analog to digital conversion of signals, vector quantization is most frequently applied directly to digital signals. This is due to the fact that vector quantization algorithms, which are complex signal processing algorithms, are used essentially in digital signal compression. A vector can be used to describe a specific pattern and, in image processing, it is made up of a set of pixels.

Vector quantization applied to bit rate reduction in data transmission or storage has been largely explored recently with focus on image data [39], [41]. This technique exploits psychovisual elements as well as statistical redundancy in the subimage wavelet coefficients obtained by the wavelet transform.

In this second part of the chapter, we present our motivations for encoding the wavelet coefficients of an image and provide a theoretical approximation of the gain yielded by vector quantization with respect to scalar quantization. The evaluation of this gain is performed under certain hypotheses regarding the probability density of the source vectors to be encoded. Since the statistics of the source vectors are generally unknown, we propose an approximation of vector quantization distortion as a function of the pdf of a subimage. As discussed in the first part of the chapter, the distribution of wavelet coefficients is not random; it corresponds to a generalized Gaussian type probability density. This approximation enables the determination of an upper bound for distortion. We also introduce the concept of the multiresolution codebook as well as an optimal bit allocation scheme in the different subimages to be encoded. This algorithm allows us to exploit the spatio-frequential aspect introduced by the transform. The vector quantizer which we developed for image compression applications is described in paragraph 2.6. It is based on the lattices defined by Conway and Sloane.

2.2. QUANTIZATION

2.2.1. Scalar Quantization

A scalar quantizer discretizes one sample at a time. It is a special case of vector quantization.

A scalar quantizer Q with L levels is entirely defined by two sets of values, $L+1$ values $x_0, x_1, \ldots, x_L$ called *decision levels*, which divide the space of real numbers, and L *reproduction values* (output values) $y_1, \ldots, y_L$ such that:

$$Q(x) = y_i \quad \text{if} \quad x_{i-1} \leq x \leq x_i \quad \text{for} \quad i = 1, 2, \ldots, L$$

An example of a scalar quantizer is given in figure 2.1. This quantizer has $L=7$ output levels. It is also known as mid tread quantizer. The quantization step is defined by the distance $[x_{i-1}, x_i]$. For example, if the input x lies between thresholds x_4 and x_5, the corresponding output value is y_5. It is generally expected that the source value is contained in the domain $[x_0, x_L]$. For values greater than x_L the reproduction value is y_L and for values smaller than x_0, the reproduction value is y_1.

Two types of noise errors occur during quantization:

- quantization noise: this is the difference between the input value and the reproduction value in the domain $[x_0, x_L]$.

- overload noise: this is the truncation effect when the signal exceeds the boundary decision thresholds x_0 and x_L and the values are quantized as y_1 or y_L.

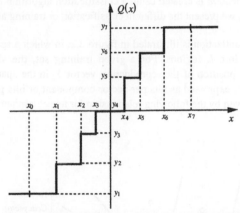

Figure 2.1.: Input-output characteristic for a scalar quantizer.

The choice of the quantization step for a scalar quantizer can be optimized on the basis of the probability distribution of the variable to be quantized. This problem has a unique solution known as the Max quantizer [39].

2.2.2. Vector Quantization

In the following X denotes a random variable of dimension n belonging to $\mathbf{R}^n$, whose joint probability density is given by $f_X(x) = f_X(x_1, x_2, \ldots, x_n)$.

A *vector quantizer* Q, of dimension n and size L, is defined as an application which associates one of the L reproduction vectors $Y_1, Y_2, \ldots, Y_L$ with a vector $X \in \mathbf{R}^n$. Thus:

$$Q : \mathbf{R}^n \rightarrow Y$$
$$X \rightarrow Q(X) = Y_i$$

(2-1)

122

where $Y = \{Y_i \in \mathbf{R}^n ; i = 1, 2, \dots, L\}$ is a set of reproduction vectors known in the literature as a *codebook*.

A vector quantizer is therefore completely defined by the knowledge of these L Y_i vectors and a partition C of the $\mathbf{R}^n$ space into L regions or classes C_i $(i = 1, 2, \dots, L)$. These regions, called *Voronoi* regions, are given by

$$C_i = \{X \in \mathbf{R}^n / Q(X) = Y_i \ \ if \ \ d(X, Y_i) \le d(X, Y_j) \ \ \forall j \ne i\}$$

$$\bigcup_i C_i = \mathbf{R}^n \ \ and \ \ C_i \cap C_j = \varnothing \ \ if \ \ i \ne j \ \ \ \forall i, j = 1, \dots, L \tag{2-2}$$

Each region contains a set of $\mathbf{R}^n$ vectors. For example, the vectors X belonging to class C_i are represented by the vector Y_i belonging to the codebook. The distance $d(X,Y)$ provides a measure of the distortion between X and Y.

Partition C for a given codebook is created using a classification algorithm and a training set. In the second part of this chapter, we present the different classification or training algorithms used.

The principle of vector quantization is illustrated in figure 2.2. in which a space of dimension $n=2$ is partitioned, or classified into L regions. For a given training set, the shape of each region is determined simply by the position of the reproduction vector Y_i in the space and by the distortion criterion used. The bit rate, expressed as bits per vector component or bits per pixel (bpp) if we are working with images, is given by the following relationship for a vector quantizer:

$$R = \frac{1}{n} \log_2 (L) \tag{2-3}$$

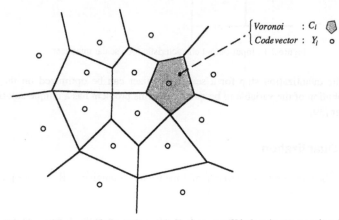

$$\begin{cases} Voronoi & : C_l \\ Code\,vector : Y_l & \circ \end{cases}$$

Figure 2.2.: Partitioning of a space of dimension $n=2$. Each vector X belonging to a region C_i is quantized by $Y_i = Q(X)$.

Relationship (2-3) enables the determination of the average number of bits per vector component needed to represent a codebook vector Y, and by the same token a vector X at the quantizer input. It is pointed out that for a given n, the bit rate depends on the number L of codebook vectors rather than on the number of bits required to encode separately the components of the vectors stored in the codebook. We present, figure 2.3, the working principle of a vector quantizer. The binary code for the index, corresponding to the codebook vector which minimizes distortion, is sent over the transmission line. Naturally, the decoder must have the same codebook as the coder, in order to reconstruct the signal. Vector quantization is thus a combination of two functions: a <u>coder</u>, which seeks the best representation $Q(X)=Y_i$ of input vector X, and supplies the index i of this latter in the codebook, and a <u>decoder</u> which uses this index to find the reproduction vector Y_i. The search for the best representation is performed by minimizing a distortion criterion $d(X,Q(X))$.

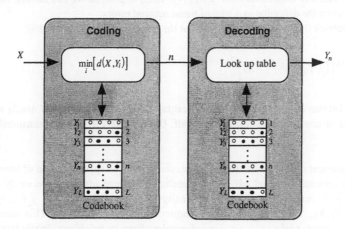

Figure 2.3.: Principle of quantization using a vector quantizer

2.2.3. Approximation of Quantizer Distortion

In 1966, Algazi [1] give a formula for the distortion of a scalar quantizer as a function of the probability density function $p_X(x)$ of the signal to quantize. Here, X is a monodimensional random variable. However, this formula is valid only for a large bit rate R in bits per sample, i.e., for a small overload noise. Thus, we have an asymptotic estimation of the distortion. This formula is given by

$$D=\frac{1}{12}2^{-2R}\left(\int_{\mathbb{R}}p_X(x)^{1/3}dx\right)^3 \tag{2-4}$$

Such a result can be obtained for a vector quantizer. In fact, for a vector quantizer, the distortion is expressed as follow. Given a random vector $X\in\mathbb{R}^n$ with joint probability density function $f_X(x)$, a codebook size L and a bit rate R in bits per sample.

The total distortion per dimension of a vector quantizer is thus given by

$$D=\frac{1}{n}E\left\{\|X-Q(X)\|^2\right\}=\frac{1}{n}\sum_{i=1}^{L}\int_{X\in C_i}\|X-Y_i\|^2 f_X(x)dx \tag{2-5}$$

In practice, the overall measure of performance is the long term sample average [39]

$$\bar{d} = \lim_{k \to \infty} \frac{1}{k} \sum_{i=1}^{k} d\big(X_i, Q(X_i)\big) \tag{2-6}$$

where $\{X_i\}$ is a sequence of vectors to be encoded and $d\big(X_i, Q(X_i)\big) = \|X_i - Q(X_i)\|^2$ is the usual L_2 norm. If the vector process is stationary and ergodic, then the above limit exists and equals the statistical expectation D defined formula (2-5) [39]. Thus, in practice we can assume that, for large k, $\bar{d} = D$ and compute the distortion as a sum of mean squared errors.

For a large codebook (large L), Zador showed that the distortion D is given by [89]:

$$D = A(n,2) 2^{-2R} \left[\int_{\mathbb{R}^n} f_X(x)^{n/(2+n)} dx \right]^{(2+n)/n} \tag{2-7}$$

where R is the bit rate in bits per sample (see paragraph 2.3.) and the values of $A(n,2)$ were tabulated by Conway and Sloane [24] for a uniform joint pdf. Unfortunately the multidimensional pdf $f_X(x)$ is unknown.

To permit a numerical evaluation of the distortion, we propose an approximation of formula (2-7) as a function of the monodimensional pdf of the quantized subimage which can be easily modeled (see paragraph 1.8.). Given a joint probability density function $f_X(x) = f_{X_1 X_2 \dots X_n}(x_1, x_2, \dots, x_n)$, we suppose that $X_1, X_2, \dots, X_n$ are independent monodimensional variables. This means that each component of a vector $x = (x_1, x_2, \dots, x_n)$ is independent from its neighbors. While this hypothesis is not really verified in image processing, it permits us to give an upper bound for the distortion D. Thus, we can write

$$f_X(x) = f_{X_1}(x_1) f_{X_2}(x_2) \dots f_{X_n}(x_n)$$

where $f_{X_i}(x_i)$ are monodimensional pdfs equal to the pdf $p_X(x)$ of the considered sub-image. In fact, we suppose that each pixel has the same pdf as its neighbors. Then, formula (2-7) of the distortion becomes [4]

$$D \leq A(n,2) 2^{-2R} \left[\prod_{i=1}^{n} \int_{\mathbb{R}} f_{X_i}(x_i)^{n/(2+n)} dx_i \right]^{(2+n)/n}$$

and can be rewritten

$$D \leq A(n,2) 2^{-2R} \left[\int_{\mathbb{R}} p_X(x)^{n/(2+n)} dx \right]^{(2+n)} \tag{2-8}$$

This theoretical upper bound approximation is verified by experiments as we can see on figure 2.4.

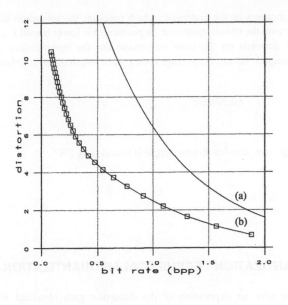

Figure 2.4.: Distortion/bit rate curves. (a) theoretical upper bound for a Laplacian law - (b) experimental results.

2.3. CODEBOOK ENTROPY

For the purposes of transmission or storage, a binary word c_i, of length b_i bits, called the index of the reproduction vector, is assigned to each output vector Y_i. Thus, vector quantization can also be seen as a combination of two functions: an encoder, which views the input vector X and generates the index of the reproduction vector specified by $Q(X)$, and a decoder, which uses this index to generate the reproduction vector Y. Let us define

$$b_i = -\log_2 p(Y_i)$$

where $p(Y_i) = prob(X \in C_i)$ is the probability of selecting the source vector Y_i during the encoding. Thus, the theoretical average binary word length is given by the formula

$$H(Y) = -\sum_{i=1}^{L} p(Y_i) \log_2 p(Y_i) \qquad \text{bits/vector} \qquad (2\text{-}9)$$

the so-called entropy measure of the codebook, which specifies the minimum bit rate necessary to achieve a distortion D with the chosen quantizer. In practice, this lower bound is never reached since $b_i \neq -\log_2 p(Y_i)$, but depends on the code we choose for the transmission (Huffman code or arithmetic code for example). Hence, the average binary word length of the codebook is given by

$$\bar{b} = \sum_{i=1}^{L} p(Y_i) b_i \qquad \text{bits/vector} \tag{2-10}$$

and the practical average rate $R = \dfrac{1}{n}\bar{b}$ in bits/sample is bounded by [39]

$$\frac{1}{n} H(Y) \leq R \leq \frac{1}{n} H(Y) + \frac{1}{n} \tag{2-11}$$

2.4. VECTOR QUANTIZATION VERSUS SCALAR QUANTIZATION

In this paragraph, we give an expression of the distortion gain obtained when doing vector quantization rather than scalar quantization in wavelet coefficient subimages. This gain is given by

$$G_{VQ} = \frac{D_{SQ}(R)}{D_{VQ}^n(R)} \tag{2-12}$$

where $D_{SQ}(R)$ is the distortion of a scalar quantizer and $D_{VQ}^n(R)$ that of a vector quantizer. The expressions for these distortions were given in paragraph 2.2.3. For purposes of comparison, the bit rate R is the same in both the scalar and vector cases. We computed an expression for the gain G_{VQ} for a subimage at a given resolution. Using the formulas of Algazi (2-4) and Zador (2-7), we can express the gain by the following formula [60], [6]:

$$G_{VQ} = \frac{1}{12 A(n,2)} \frac{\left[\int_{\mathbf{R}} p_X(x)^{1/3} dx \right]^3}{\left[\int_{\mathbf{R}^n} f_Z(z)^{n/(2+n)} dz \right]^{(2+n)/n}} \tag{2-13}$$

where X is a monodimensional random variable and Z a multidimensional random variable.

Note that G_{VQ} does not depend on the bit rate R assigned to the subimage. However, because of the approximations used in the development of the distortion formulas (2-4) and (2-7), the above formula applies only when R is large, and thus provides an asymptotic indication only. Since the joint probability density is unknown, numerical evaluation of the gain calls for use of the approximation (2-8) in paragraph 2.2.3. We thus obtain the <u>minimum theoretical gain</u> which can be achieved. This gain is expressed as follows:

$$G_{VQ} \geq \frac{1}{12\,A(n,2)} \frac{\left[\int_{\mathbf{R}} p_X(x)^{1/3}\,dx\right]^3}{\left[\int_{\mathbf{R}} p_X(x)^{n/(2+n)}\,dx\right]^{(2+n)}} \tag{2-14}$$

where $p_X(x)$ corresponds to the pdf of the subimage in which the gain is evaluated.

Table (2-1) summarizes the different expressions of G_{VQ} using different approximate models for the subimage pdf. Based on these different approximations, we plotted in figure 2.5. the theoretical G_{VQ} curves as a function of vector size n. These curves give an idea of the minimum attainable gain using vector quantization.

As can be seen in figure 2.5., the generalized Gaussian law gives the best results in terms of minimum achievable gain. These results were obtained for independent vector components. It is thus clear that vector quantization is a more powerful tool than scalar quantization, even when this latter is optimum, particularly in association with the wavelet transform.

In summary Vector Quantization performs better for coding wavelet coefficients.

Model	$p_X(x)$	Gain G_{VQ}
Laplacian	$\dfrac{1}{\sigma\sqrt{2}}\exp\left(-\dfrac{\sqrt{2}|x|}{\sigma}\right)$	$\dfrac{9}{4\,A(n,2)\left(\dfrac{2+n}{n}\right)^{(2+n)}}$
Gaussian	$\dfrac{1}{\sigma\sqrt{2\pi}}\exp\left(-\dfrac{x^2}{2\sigma^2}\right)$	$\dfrac{\sqrt{3}}{4\,A(n,2)\left(\dfrac{2+n}{n}\right)^{(2+n)/2}}$
Generalized Gaussian $(\alpha=1/2)$	$\dfrac{1}{2\sigma}\sqrt{30}\,\exp\left(-\sqrt{\dfrac{2}{\sigma}}\sqrt{30}|x|\right)$	$\dfrac{243}{4\,A(n,2)\left(\dfrac{2+n}{n}\right)^{2(2+n)}}$

Table 2.1.: Theoretical gains G_{VQ} for different approximation models of the subimage pdfs. The pdf is supposed to be zero mean.

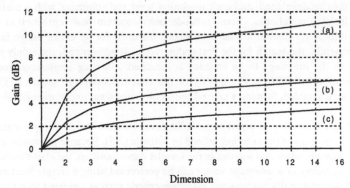

Figure 2.5.: Theoretical minimal gain G_{VQ}, expressed in dB, versus vector size n, for three different pdf models. (a) Generalized Gaussian law ($\alpha=0.5$) - (b) Laplacian law ($\alpha=1$) - (c) Gaussian law ($\alpha=2$). The values of $A(n,2)$ correspond to the conjectured lower bound of distortion computed by Conway & Sloane [24].

2.5. CODEBOOK DESIGN

2.5.1. Introduction

In the development of a vector quantizer, the main goal is to find a codebook which minimizes a distortion criterion. The achievement of this codebook yields a globally optimal quantizer. In practice, however, this is rarely attained since codebook generation algorithms lead to locally optimal quantizers. Indeed, for a given sequence of vectors to be encoded and a given codebok size, the most difficult task is to construct a codebook containing the best collection of reproduction vectors and which efficiently represents the broad variety of source vectors to be encoded.

Numerous codebook generation algorithms have been proposed in the literature. A synthesis of these different methods is beyond the scope of the present work; it is simply recalled in the third part of this paragraph that the work of Linde, Buzo, and Gray led to the development of a widely known iterative algorithm called the LBG [54], used extensively in image processing.

In most vector quantization applications, there is no *a priori* knowledge of the source vectors to be encoded; their probability density function $f_X(x)$ is unknown. For this reason, most codebook generation algorithms use a training set. This training set contains a sample of each of the many source vectors to be encoded.

However, the development of recent algebraic compression techniques such as lattice vector quantization enables these training sets to be avoided. In this case, the codebook is no longer computed by partitioning, but rather defined *a priori* as a set of regularly distributed vectors. The construction of these quantizers is described paragraph 2.6.

2.5.2. Multiresolution Codebook

In the preceding paragraphs, we explained the advantages of associating vector quantization and the wavelet transform. Indeed, the decomposition of an image into several resolution levels and for different edge directions (dyadic case), yields subimages whose statistical characteristics are much easier to work with, than those of the original image. Codebooks specifically adapted to each type of subimage can thus be developed, and each resolution level and subimage with a preferential edge direction has its own codebook. These "subcodebooks" contain few reproduction vectors and, moreover, exhibits lower distortion than a codebook obtained for an entire image in the original domain. Consequently, the search for the best coding vector is accelerated, since only the codebook corresponding to the subimage to be encoded is searched, allowing implementation of parallel processing schemes. In addition, the quality of the encoded image is superior. These various subcodebooks are assembled, forming a general codebook called a *multiresolution codebook*, depicted figure 2.6.

Each codebook corresponding to a given resolution level contains the most representative wavelet coefficient forms, in terms of the criterion defined by formula (2-5). Because the spatial and frequency aspects of the image are taken into account in the wavelet decomposition, the classification and search involved in the encoding of a subimage vector can be performed using a simple least mean squares-type criterion. Computationally burdensome, distortion criteria such as weighted least mean squares or other measurements based on perceptual factors can thus be avoided.

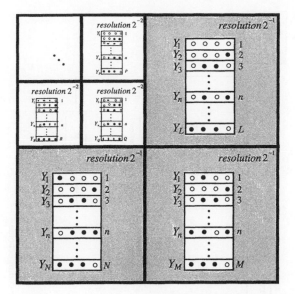

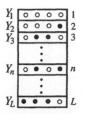

codebook with size L

Figure 2.6.: Multiresolution codebook

Figure 2-7 shows a general image coding scheme using the wavelet transform and vector quantization [60], [6]. The low-frequency image is quantized using a scalar quantizer or DPCM (Differential Pulse Code Modulation) [70] and transmitted as wavelet coefficient subimages. The problem of entropy coding was explored using arithmetic and Huffman-type methods [44], [75]. Experimental results using these methods [21] were observed to be within 10% of those estimated by Shannon's minimum boundary. Transmission errors due to noisy channels are beyond the scope of the present work. Readers interested in entropy coding and the modeling of noise in transmission channels are referred to [87], [21].

2.5.3. Learning Methods: LBG, ECVQ

2.5.3.1. LBG Algorithm

This classification algorithm is based on the simple observation that Lloyd's algorithm (1957) developed for scalar quantization [56] is also applicable to vectors, sampled distributions, and a broad variety of distortion criteria. The principle consists in iteratively modifying a codebook, thus generalizing Lloyd's algorithm. This generalization, known as "k-means" is often referred to as the LBG [54] algorithm since it was developed by Linde, Buzo, and Gray for vector quantizers in 1980. Actually, the name LBG is more appropriate for a variation of Lloyd's algorithm which uses the vector splitting technique in order to construct the codebook.

The LBG algorithm was designed to perform a classification operation on a *training set* for a given initial codebook. This training set is made up of source vectors, whose probability density is unknown, to be encoded. There are two steps:

130

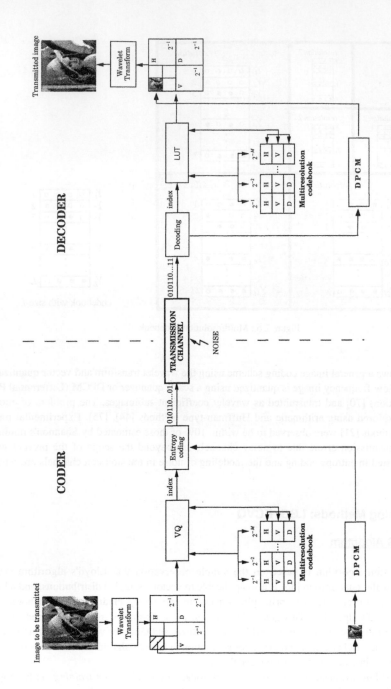

Figure 2.7.: General image coding scheme

- The first step involves a <u>classification</u> enabling each vector in the training set to be associated with its most highly representative reproduction vector in the initial codebook. This involves partitioning the space containing the vectors to be encoded.

- The second step is <u>optimization</u> which, by computing the center of gravity (centroid) of each class of vectors created in the preceding step, ideally matches the codebook vector to the class it represents.

The algorithm is reiterated with this new codebook, in order to obtain a new partitioning. It converges to a stable position, improving the given distortion criterion at each iteration. Indeed, in Lloyd's algorithm, after each iteration, the mean distortion must either be reduced or remain the same [37], [38]. The choice of the initial codebook (created by splitting) has a major influence on the determination of the local minimum the algorithm reaches (the global minimum corresponds to a possible initial codebook).

The complexity of this algorithm is proportional to 2^{nR}, n being the vector size and R the bit rate associated with the codebook $R = (1/n)\log_2(L)$.

2.5.3.2. <u>Entropy-Constrained Vector Quantization (ECVQ)</u>

For a fixed vector length and distortion $D_{VQ}^n(R)$, as seen in the first part of this chapter, if the indices of the codebook vectors are coded entropically, the corresponding bit rate is lower than if a natural (fixed code length) encoding method is used [22]. The ECVQ (Entropy-Constrained Vector Quantization) algorithm, which is a generalization of the LBG algorithm, enables the creation of vector quantizers with minimum distortion by imposing a constraint on the entropy. The problem then consists in finding a set of reproduction vectors Y which minimizes the following functional

$$J = D_{VQ}^n + \mu H(Y)$$

This ECVQ algorithm is optimal, but extremely <u>costly</u> in terms of CPU time.

In 1979, Gersho [37] conjectured that the codebook obtained by use of the optimal ECVQ algorithm should take the form of a lattice. A lattice in the R^n space is made up of independent vectors which cover the entire space. The components of these vectors are obtained by linear combinations of integer numbers. The principle of lattice vector quantizers is presented in the following paragraph.

2.6. LATTICE VECTOR QUANTIZATION (LVQ)

2.6.1. Introduction

In this paragraph, we describe the construction of quantizers based on lattices (sets of regularly spaced points). The codebook is no longer computed using training sets, as in the case of the previously described LBG and ECVQ algorithms, but rather defined *a priori* as a set of points (vectors) regularly distributed throughout the space. Our study will begin with a simple example of a lattice, the Z^n lattice, which will enable us to introduce the essential concepts of lattice vector quantization.

The Z^n lattice is defined as a set of points whose coordinates are integers. It is also called the set of integers.

$$Z^n = \left\{ Y = (y_1, y_2, ..., y_n) \in R^n \,/\, y_i \in Z \right\} \tag{2-15}$$

In two dimensions, the lattice points are evenly distributed on a square grid, and in 3 dimensions, on a cubic lattice (see figures 2.12. and 2.13.). In order to build a vector quantizer based on a Z^n lattice, we recall the definitions given in paragraph 2.2.2.:

-   The codebook is defined as a finite subset of points in Z^n. We select the points contained in a *ball* of radius m for the L_p metric, centered on the origin. These points are said to be of "energy" less than m^p.

$$Y = \left\{ Y = (y_1, ..., y_n) \in Z^n \,/\, \sum_{i=1}^{n} |y_i|^p \leq m^p \right\} \tag{2-16}$$

These lattice points make up the points in the codebook. The shape of the classes associated with each point is defined implicitly by the choice of lattice. For the dimension $n=1$, we find straight segments $[-1/2, 1/2[$ of the scalar quantizer whose interval is unity. For $n=2$, we find squares, and $n=3$, cubes (see figures 2.12. and 2.13.). The definition of Voronoi classes or regions is given further on in this paragraph.

-   The quantization algorithm is elementary. The representation of a source point is the closest integer. To quantize a vector, the search is for the closest integer to each of its components. For example, the source vector $X = (0.9, -1.2)$ is quantized by the reproduction vector $Y_i = (1, -1)$.

-   An index i and a binary code b_i are assigned to each lattice vector Y_i for transmission and decoding at the receiver.

2.6.2. Description of the lattices used

2.6.2.1. Definition

An n-dimensional lattice Λ is defined as a set of points Y belonging to R^n such that

$$\Lambda = \left\{ Y \in R^n \,/\, \exists (u_1, ..., u_n) \in Z^n, \quad Y = \sum_{i=1}^{n} u_i a_i \right\} \tag{2-17}$$

where $a_1, a_2, ..., a_n$ are linearly independent vectors belonging to an m-dimensional space R^m where $m \geq n$. The generating matrix G is a matrix whose rows are the basic vectors of the lattice: $G^t = [a_1^t ... a_n^t]$. In particular, the determinant of the lattice gives the volume of the elementary Voronoi:

$$vol(\Lambda) = \det{}_\Lambda = \left| \det GG^t \right|^{1/2}$$

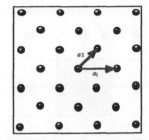

Figure 2.8.: Two-dimensional lattice generated by the vectors a_1, a_2.

Lattices can be seen as a solution to problems involving spheres packed in a space or covering a space [27], where the centers of the spheres define the points in the lattice. In the first case, for a given size, the problem is to determine the most densely packed configuration (for example, in 2 dimensions, the hexagonal lattice is the most dense). Let ρ denote the packing radius. This radius is that of the spheres packed without overlap. The second problem is dual. It involves covering the space entirely with spheres, with the smallest overlap possible between spheres. The covering radius R corresponds to the maximum distance between one point in the space and its closest point(s) in the lattice.

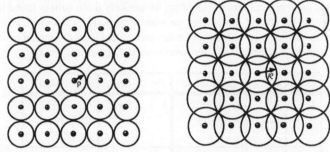

Figure 2.9.: Spheres packed in a space and covering a space. The packing radius ρ and the covering radius R are defined respectively.

The reproduction vectors of a lattice vector quantizer correspond to the regularly spaced lattice points, as defined by relationship (2-17). The predefined distribution of vectors yields a codebook whose structure is periodic and ordered, enabling implementation of fast and simple quantization algorithms. A Voronoi class or region C_i is the set of points lying closer to Y_i than to any other point in the lattice. The definition of the lattice implies the definition of the associated Voronoi. A Voronoi corresponds to a convex region of $\mathbf{R}^n$, surrounded by a finite number of hyperplanes. Hyperplanes lie midway between the reproduction vector under consideration and its closest neighbors in the lattice (see figure 2.10). Using this definition, and assuming that the probability density is locally uniform in each polytope, the lattice points are indeed the centroids of the Voronoi regions.

When a lattice is used for quantization of a uniform source, the normed mean square error is the same for all classes C_i. In that case, the performance of the quantizer depends solely on the geometry of the class C_0 of the lattice. Thus the error per dimension is given by the normalized second-order moment of inertia $G_n(\Lambda)$ given below:

134

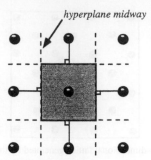

Figure 2.10.: The convex shaded area defined by hyperplanes represents a Voronoi region.

$$G_n(\Lambda) = \frac{1}{n} \frac{\int_{C_0} \|Y\|^2 \, dY}{\left(\int_{C_0} dY\right)^{(n+2)/n}} \qquad (2\text{-}18)$$

Use of the normalized value enables comparison of quantizer performance regardless of size.

Definition (2-17) is quite general. For quantization, we use only those lattices found in the literature, which are optimal for the size of the vectors we wish to quantize. Conway and Sloane [25], [27] studied the properties of the most well known lattices and drafted a list of optimal quantizers per size. Table 2.2. below summarizes these results for dimensions 1, 4, 8, and 16.

dim	lattice	some values	dim	lattice	some values
1	Z^1	$G_1 = 0{,}0833$ $R = \sqrt{n}/2$ $\rho = 1/2$	8	E_8	$G_8 = 0{,}0717$ $R = \rho\sqrt{2} = 1$ $\rho = 1/\sqrt{2}$
4	D_4	$G_4 = 0{,}0766$ $R = \rho\sqrt{2} = 1$ $\rho = 1/\sqrt{2}$	16	Λ_{16}	$G_{16} = 0{,}0683$ $R = \rho\sqrt{3}$ $\rho = 1$

Table 2.2.: Optimal lattices for dimensions 1, 4, 8, and 16. A number of characteristic lattice values are recalled: the normalized second-order moment of inertia $G_n(\Lambda)$, the covering radius R and the packing radius ρ.

2.6.2.2. Z^n Lattice

The Z^n lattice is the simplest of all lattices. It is comprised of all points in R^n whose coordinates are integers. This lattice is also called a cubic lattice.

$$Z^n = \left\{ Y = (y_1, y_2, \ldots, y_n) \, / \, y_i \in Z \right\} \qquad (2\text{-}19)$$

Several Z^n lattices are depicted in the following paragraph. They correspond to $n=1$, 2, and 3.

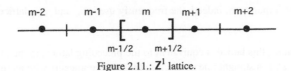

Figure 2.11.: Z^1 lattice.

The black points with integer coordinates make up the Z^1 lattice. This lattice is the straight line joining the set of integers. In analogy to scalar quantization, the decision intervals of Z^1 are of the form $[m-1/2, m+1/2[\ m \in Z$.

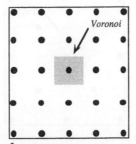

Figure 2.12.: The Z^2 lattice = set of points whose coordinates are integers. The Voronoi region is a *square*.

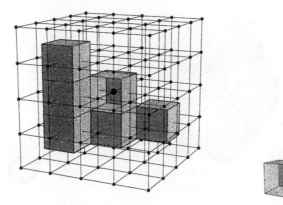

Figure 2.13.: 3D cubic lattice. The Voronoi region is a *cube*.

2.6.2.3. D_n Lattice

The D_n lattice corresponds to those points in the cubic lattice Z^n for which the sum of their coordinates is even.

$$D_n = \left\{ Y = (y_1, ..., y_n) \in Z^n \ / \ \sum_{i=1}^{n} y_i = 0 \ \mathrm{mod}(2) \right\} \qquad n \geq 3 \qquad (2\text{-}20)$$

The D_n lattice is of importance; indeed, the frequently used E_n and Λ_{16} lattices are built from this lattice.

For 1 and 2 dimensions, this lattice is equivalent to the preceding lattice, to the exception of a dilation and rotation factor. D_1 is a straight line connecting the relative integers with an interval of $\sqrt{2}$ instead of 1. D_2 can be compared to a checkerboard, in which only one color is taken into consideration; in this case, the Voronoi is a losange. The D_3 lattice is depicted in figure 2-15. For this lattice, the Voronoi regions are rhombic dodecahedra.

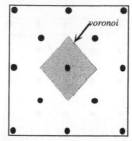

Figure 2.14.: This lattice is usually compared to a checkerboard in which only one color is taken into consideration. The Voronoi is a *losange*.

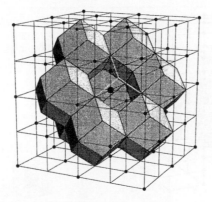

Figure 2.15.: 3D cubic lattice with centered facets (D_3). The lattice points are identified by black dots. On the right, we have shown the Voronoi centered on the origin, with one transparent facet: *rhombic dodecahedron*.

In the fourth dimension, the vector quantizer built from this lattice is optimal for a uniform input signal distribution. As seen further, the D_4 lattice can be obtained by a construction A applied to the parity check code. The construction of lattices from binary codes is described in paragraph 2.6.5.

2.6.2.4. E_8 Lattice

The E_8 lattice is defined by the following relationship:

$$E_8 = D_8 \cup \left(\frac{1}{2}[1] + D_8\right) \tag{2-21}$$

where [1] is the vector whose components are all equal to 1.

The E_8 lattice is the union of two D_8 lattices, one shifted by 1/2 along each component. This lattice contains the points in the cubic lattice with an even coordinate sum as well as the shifted points in the cubic lattice whose coordinates also form an even sum. This lattice can be obtained by applying construction A (Cf. paragraph 2.6.5.) to the H_8 Hamming code of length 8 or by applying construction B to the code $\mathcal{R} = \left\{0^8, 1^8\right\}$.

2.6.2.5. Λ_{16} Lattice

The Λ_{16} lattice is defined by the following relationship:

$$\Lambda_{16} = \bigcup_{i=0}^{31} \left(r_i + 2D_{16}\right) \tag{2-22}$$

where the translation vectors r_i correspond to the rows (or columns) of a Sylvester-type Hadamard matrix $\tilde{H}_{16}$ with the transformation $\{-1,1\} \rightarrow \{1,0\}$, and to the rows (or columns) of the complemented matrix $\bar{\tilde{H}}_{16}$.
Actually, Λ_{16} is the union of 32 D_{16} lattices shifted by the translation vectors r_i.

2.6.3. Fast Quantization Algorithms

The principle of these algorithms is to represent a source vector $X \in \mathbf{R}^n$ by a vector Y belonging to the given lattice [27]. In the following, it is assumed that the lattices are infinite, i.e., that there is no overload noise.

2.6.3.1. Quantization in a Z^n lattice

This type of quantization was addressed in the introduction. Here, it is formalized in order to introduce quantization in more complex lattices such as the D_n.

Let $f(x)$ be a function which, for any real number x, associates its closest integer. To vector quantize in a Z^n lattice, $f(x)$ is applied to each vector component.

$$Y = f(X) = \left[f(x_1), \ldots, f(x_n)\right] \qquad X \in \mathbf{R}^n \quad \text{and} \quad Y \in Z^n \tag{2-23}$$

Actually, this amounts to performing scalar quantization with an interval unity on each component of the source vector. There is no overload noise since the edge effect is not taken into account with the codebook. This problem will be addressed later by truncating the lattice.

2.6.3.2. Quantization in a D_n lattice

Algorithm

Conway and Sloane developed fast quantization algorithms for vector quantizers based on the D_n lattice [25].
Let $f(x)$ be the closest integer to x and $w(x)$ the wrong way from x defined by

$$w(x) = f(x) + \text{sign}(x - f(x)) \quad \text{with} \quad \begin{cases} \text{sign}(z) = 1 & \text{if } z \geq 0 \\ \text{sign}(z) = -1 & \text{else if} \end{cases}$$

Let $X = (x_1, \ldots, x_n)$ be the source vector to be quantized and $\delta = x - f(x)$ the quantization error. The search for the lattice point of D_n the closest to $X \in \mathbf{R}^n$ entails the following steps:

① Compute $f(X) = (f(x_1), f(x_2), \ldots, f(x_n))$ the point in $\mathbf{Z}^n$ the closest to X.

② Compute $g(X) = (f(x_1), \ldots, w(x_i), \ldots, f(x_n))$ where x_i is the component for which the quantization error δ_i is the greatest. For this coordinate indexed i, $f(x_i)$ is replaced by $w(x_i)$ in order to form $g(X)$.

③ These two vectors differ by a single component and the sum of their components differs by one unit. If one has an even sum, the other has an odd sum. The closest point in D_n is that with the even sum; hence it is selected as reproduction vector.

Complexity

Unlike LBG algorithms, there is no need to compute the norm to search for the closest reproduction vector among all those in the codebook. In addition, the computational cost does not depend on the size of the codebook.
Quantization in a D_n lattice involves:

$f(X)$: n round-offs
n-1 sums
1 test for parity

$g(X)$: n differences (the δ_i)
1 search for the $\max(\text{abs}(\delta_i))$
1 round-off (computation of $w(x_i)$)

i.e., at least $2n$ operations (if the sum of the components of $f(X)$ is even) and $3n+2$ operations at most (if the sum of the components is odd).

In comparison, the computational load for vector quantization in the case of a complete search in a codebook of L vectors involves:

> nL differences
> nL squares
> $(n-1)L$ sums

or a total of $(3n-1)L$ operations, depending on the size L of the codebook.

2.6.3.3. Extension to other lattices

A procedure for finding the point closest to X in a lattice Λ can easily be extended to find the closest point in a "coset"[1] $r+\Lambda$ or in a union of cosets $\bigcup_i (r_i + \Lambda)$. If $Q_\Lambda(X)$ is the closest point to X in the lattice Λ, then $Y = Q_\Lambda(X-r)+r$ is the closest point to X in the shifted lattice $r+\Lambda$. For a union of cosets, all of the $Y_i = Q_\Lambda(X-r_i)+r_i$ must be computed, and the closest to X in terms of minimal distortion must be selected [27].

Since E_8 is a union of two D_8 cosets and Λ_{16} a union of 32 D_{16} cosets, we can then use the same coding procedure as previously for D_n in order to find the closest point to X in E_8 and Λ_{16}, with some additional computations for distance.

Quantization in the E_8 lattice

It is recalled that the E_8 lattice is the union of two D_8 cosets, one shifted with respect to the other. Quantization in this lattice involves:

- a search for the closest neighbor to X in the D_8 lattice
- a search for the closest neighbor to X in the $(1/2)[\mathbf{1}] + D_8$ lattice
- selection of the better of the two (in terms of the L_2 norm).

Overall quantization complexity is the sum of the following:

① quantization complexity in the D_8 lattice
② quantization complexity in the shifted D_8 lattice (shift, quantization in D_8, inverse shift)
③ two norms to be computed $\left\| Y_{D_8} - X \right\|$ and $\left\| Y_{(1/2)[\mathbf{1}]+D_8} - X \right\|$, and the search for the minimum of the two norms

The overall complexity is independent of the size of the codebook. It is lesser than in the case of a complete search.

Quantization in the Λ_{16} lattice

The Λ_{16} lattice is the union of 32 D_{16} shifted lattices. Quantization thus involves:

- a search for the closest neighbor to X in each of the shifted lattices
- selection of the closest (in terms of the L_2 norm) of the 32 possible representations

---

[1] Shifted lattice.

Here again, complexity is not affected by the size of the codebook, and is lesser than in the case of a complete search.

2.6.4. Construction of the Codebook

The lattices defined in paragraph 2.6.2. have an infinite number of points. Thus, in order to limit the quantization bit rate, only a portion of the lattice points is used during quantization. A codebook containing a finite number of points must therefore be defined, based on an infinite lattice.

The shape of the truncation envelope is determined by the statistical nature of the source vectors to be encoded. In practice, Gaussian or Laplacian laws are used because they provide a good approximation of the statistical properties of the wavelet coefficient subimages in quantization algorithms.

Fischer [30] conducted a study on Laplacian sources for which the equiprobability curves are pyramidal. In this approach, the codebook is made up of the points in a pyramid. The source vector is encoded by the pyramid radius and by the index of the vector on the pyramid. A more general method proposed by Jeong and Gibson [47] consists in taking a lattice such that the point density decreases as the distance from the origin increases. However, because there is no equivalent of theta functions (series enabling the counting of vectors in the L_2 norm [see 2.6.5.1.]) for the L_1 norm, these authors were obliged to use the cubic lattice. The introduction of Nu functions (equivalent of theta functions for the L_1 norm) enables us to extend quantization to lattices of greater density than the cubic lattice.

Once the truncation has been selected, the coding of vectors in the codebook can be performed, either by an entropic code which takes into account their probability of appearing during the coding [9],[15], or by a product code [65]. The product code is made up of the code corresponding to the energy of the quantized vector, i.e., the sphere (or pyramid) on which it is located, as well as the code corresponding to its position on the sphere (or pyramid). In the second case, the choice of the truncation envelope is more crucial if we desire a product code close to the entropic code. Nonetheless, the solution using the product code is more elegant since it is more realistic and rugged since it does not require construction of an entropic code such as Huffman or arithmetic codes.

Let X_i be a random variable whose probability density is $p_{X_i}(x_i)$.

If the x_i are arranged in vectors X of length n, then the probability density becomes, assuming that the variables are iid (independent and identically distributed - see paragraph 2.2.3.):

$$f_X(x) = \prod_{i=1}^{n} p_{X_i}(x_i) \tag{2-24}$$

Truncation is related to the statistical characteristics of the source to be quantized. The envelope of the codebook is a hypersurface such that the vectors have the same probability. This envelope is defined as follows:

$$S_T = \left\{ Y : f_Y(y) = \text{proba}\left(\|Y\|_p = m \right) \right\} \quad \text{where } m \in \mathbf{R}^+ \text{ is a constant} \tag{2-25}$$

The codebook vectors are the lattice points contained in a *ball* of radius m centered on the origin. These points are those whose "energy" is less than m^p.

$$Y = \left\{ Y = (y_1, \ldots, y_2) \in \Lambda / \sum_{i=1}^{n} |y_i|^p \leq m^p \right\}$$

This relationship also defines the norm L_p.

2.6.4.1. Spherical Truncation

We assume a source vector whose probability density is a multidimensional Gaussian.

$$f_X(x) = \frac{1}{\left(\sigma \sqrt{2\pi} \right)^n} \exp\left(-\frac{1}{2\sigma^2} \sum_{i=1}^{n} x_i^2 \right) \tag{2-26}$$

For this type of distribution, vectors X with the same probability of occurrence are located on surfaces of equal energy. Indeed, density is constant when the argument of the exponential is constant, i.e.,

$$\sum_{i=1}^{n} x_i^2 = \|X\|^2 = m^2 \tag{2-27}$$

This equation defines a surface of energy m^2 and also corresponds to the Euclidean or L_2 norm. For $n=3$, this surface is a <u>sphere</u>. The term hypersphere will be used for higher dimensions.

For this type of source, the lattice is truncated by a hypersphere. The set of points making up the codebook is defined by:

$$S(n,m) = \left\{ Y = (y_1, \ldots, y_n) \in \Lambda / \sum_{i=1}^{n} y_i^2 \leq m^2 \right\} \tag{2-28}$$

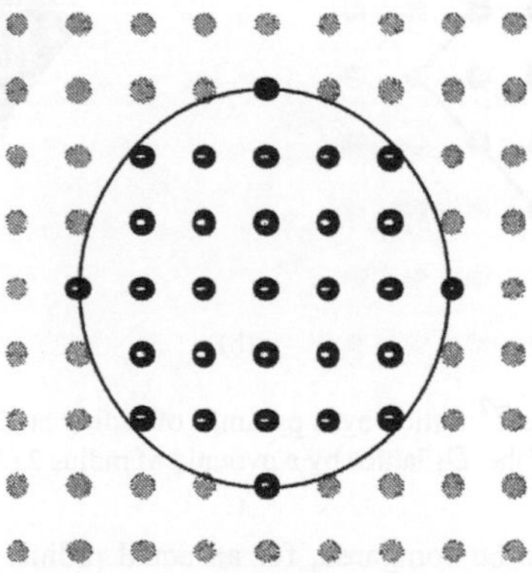

Figure 2.16.: Truncation of the Z^2 lattice by a sphere. The codebook is made up of the dark points. The truncation energy is $m^2 = 9$ (29 vectors in the truncated lattice).

142

2.6.4.2. Pyramidal Truncation

When the source distribution obeys a Laplacian law, then

$$f_X(x) = \frac{1}{\left(\sigma\sqrt{2}\right)^n} \, \exp\left(-\frac{\sqrt{2}}{\sigma} \sum_{i=1}^{n} |x_i|\right)$$ (2-29)

Vectors X with the same probability of appearing correspond to having:

$$\sum_{i=1}^{n} |x_i| = \|X\|_1 = m$$ (2-30)

for $n=3$, these vectors are on a pyramidal surface (see figure 2.17.). This results is also found in [30]. However, unlike Fischer, the codebook here is made up of all of the points contained in the hypersphere, not simply those on the surface (or the union of several surfaces).
The set of lattice points comprising the codebook is defined by formula (2-31). It encompasses the points whose energy in the L_1 norm is less than or equal to m.

$$S(n,m) = \left\{ Y = (y_1, \dots, y_n) \in \Lambda \, / \sum_{i=1}^{n} |y_i| \le m \right\}$$ (2-31)

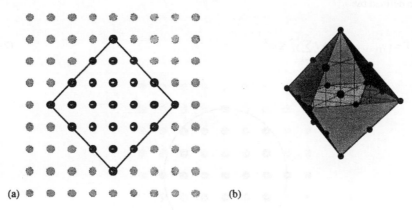

(a)   (b)

Figure 2.17.: (a) Truncation of the Z^2 lattice by a pyramid of radius $m=3$ (25 reproduction vectors in the truncated lattice) - (b) Truncation of the D_3 lattice by a pyramid of radius 2 (19 reproduction vectors).

The two types of truncation can be compared, for an equal radius. In this case, the pyramids are inscribed in the spheres.

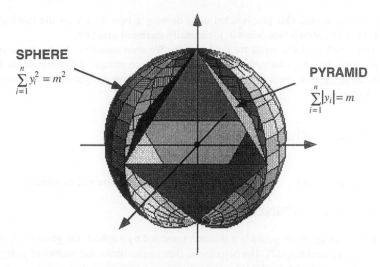

SPHERE
$$\sum_{i=1}^{n} y_i^2 = m^2$$

PYRAMID
$$\sum_{i=1}^{n} |y_i| = m$$

Figure 2.18.: Comparison between pyramids and spheres of equal radius.

2.6.4.3. Other Types of Truncation

Other types of truncation can be defined for different distributions of vectors in space. For example, the distribution of vectors in a subimage of wavelet coefficients yielded by a dyadic transform is neither spherical nor pyramidal. In this case, the coefficients are distributed throughout an oriented ellipse. This situation leads us to develop new statistical models and truncations adapted to this type of signal. These models are not discussed in the present work. Interested readers are referred to [16], [65].

2.6.5. Enumeration of Points in a Codebook

The size of the codebook depends on the number of lattice points in the truncated structure. These points are distributed on concentric balls in a manner "comparable" to Russian dolls. Consequently, the overall size of the codebook is given by the sum of all of the points on each surface contained in the structure.

To enumerate these points, one approach is to use the generating series introduced by Conway and Sloane as part of their work on lattices. For a given lattice Λ for example, the number N_m of points at distance m^2 from the origin (metric in the L_2 norm) is given by the theta series defined by [27]:

$$\Theta_\Lambda(q) = \sum_{\gamma \in \Lambda} q^{\|\gamma\|^2} = \sum_{m=0}^{+\infty} N_m q^{m^2} \quad \text{where} \quad q = e^{i\pi z} \tag{2-32}$$

Theta functions for the lattices we use have been explored and computed by many theoreticians, Conway and Sloane in particular, for a variety of applications. However, no function of this type is available for the L_1 norm. This problem led us to develop a new theory for the fabrication of Nu series in order to enumerate lattice points in pyramidally truncated structures.

We first briefly recall the theta series for the L_2 norm. We then introduce the Nu series and present results for the usual lattices. Lastly, the two types of series are compared in terms of the number of points contained in a ball of given radius. The ball is a generic term designating any convex region. It is defined as follows:

$$B_p(n,m) = \left\{ X = (x_1, \dots, x_n) \in \mathbf{R}^n \,/\, \sum_{i=1}^{n} |x_i|^p \leq m^p \right\}$$
(2-33)

where p indicates the norm L_p. This term encompasses both the sphere and the pyramid.

2.6.5.1. Theta Series (spheres)

For a codebook made up of the points in a lattice Λ truncated by a sphere, the generating Θ series are used to determine its cardinal [27]. The objective is then to enumerate the vectors of energy m^2, i.e., the vectors located on the surface of a sphere of radius m.

In formula (2-32), N_m represents the number of vectors on the circle of radius m; it is the coefficient of the term q^{m^2}. The number of points contained within a sphere of radius m (surface of energy m^2) is thus $N_T = \sum_{i=0}^{m} N_i$. By the same token, N_T represents the coefficient of the term q^{m^2} in the modified series $\Theta_\Lambda(q)/(1-q)$.

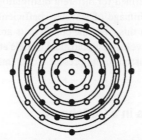

Figure 2.19.: Enumeration of points belonging to a sphere of radius 4 for the $\mathbf{Z}^2$ lattice. We count the number of vectors on the circles of energy 0 (1 vector, white), of energy 1 (4 vectors, black), of energy 2 (4 vectors, white),..., and of energy 16 (4 vectors, black). The sum of these vectors is 49 (vectors contained within a sphere of radius 4, energy = 16). See the table in Appendix B.

In order to easily construct the Θ series defined in paragraph 2.6.5, we introduce Jacobi's theta functions:

$$\theta_2(z) = \sum_{m=-\infty}^{+\infty} q^{(m+1/2)^2} = 2q^{1/4} + 2q^{9/4} + 2q^{25/4} + \dots$$

$$\theta_3(z) = \sum_{m=-\infty}^{+\infty} q^{m^2} = 1 + 2q + 2q^4 + 2q^9 + \dots$$

$$\theta_4(z) = \sum_{m=-\infty}^{+\infty} (-q)^{m^2} = 1 - 2q + 2q^4 - 2q^9 + \ \dots$$

The theta series for lattices are directly defined in terms of Jacobi's theta functions:

$$\Theta_{Z^n}(z) = \left[\theta_3(z)\right]^n$$

$$\Theta_{D_n}(z) = \frac{1}{2}\left[\theta_3(z)^n + \theta_4(z)^n\right]$$

$$\Theta_{E_8}(z) = \frac{1}{2}\left[\theta_2(z)^8 + \theta_3(z)^8 + \theta_4(z)^8\right] = 1 + 240q^2 + 2160q^4 + \dots$$

$$\Theta_{A_{16}}(z) = \frac{1}{2}\left\{\theta_2(2z)^{16} + \theta_3(2z)^{16} + \theta_4(2z)^{16} + 30\theta_2(2z)^8\theta_3(2z)^8\right\} = 1 + 4320q^4 + 61440q^6 + \dots$$

Tabulated values for these functions are given in Appendix B.

2.6.5.2. Nu Series (pyramids)

We now enumerate the points in a hyperpyramid or $L_1 - ball$. For a given Λ lattice, we define its Nu function by the relationship:

$$\nu_\Lambda(z) = \sum_{Y \in \Lambda} z^{\|Y\|_1} = \sum_{m=0}^{+\infty}\left[z^m\right]\left\{Y \in \Lambda / \|Y\|_1 = m\right\} \tag{2-34}$$

The similarities and differences with respect to the definition of the theta series (below) can be observed.

$$\Theta_\Lambda(q) = \sum_{Y \in \Lambda} q^{\|Y\|^2} = \sum_{m=0}^{+\infty}\left[q^{m^2}\right]\left\{Y \in \Lambda / \|Y\|^2 = m^2\right\} \tag{2-35}$$

The coefficient of z^m indicates the number of points located <u>on the ball</u> of energy m.

For certain lattices, intuitive reasoning is sufficient to enumerate the points in the codebook. However, when the lattices become dense, more theoretical binary code concepts are called for. Indeed, algebraic lattices can be constructed from error correcting codes. We use the A and B stacking methods of construction based on binary codes, proposed in the 1970s by Leech and Sloane. The words in the binary code are points on the cube of size unity. The type of construction determines how the pattern is repeated to form an infinite lattice.

The definition of the Nu function is based on the weight enumerator polynomial which characterizes the code and differs according to the construction chosen. For the E_8 lattice, for example, constructions A and B do not lead to the same Nu function although they do lead to the same theta function. Unlike the Euclidean norm, the L_1 norm depends on the choice of orthogonal basis. The type of construction must therefore be chosen such that the number of points in the codebook is minimal for a given radius.

Nu Series for Simple Lattices

For the **Z** lattice, which corresponds to a uniform quantizer, the *Nu* function is a geometrical series:

$$v_Z(z) = 1 + 2\sum_{m=1}^{+\infty} z^m = 1 + \frac{2z}{1-z} = \frac{1+z}{1-z} = 1 + 2z + 2z^2 + 2z^3 + \ldots \qquad (2\text{-}36)$$

For this lattice, there is one point of energy 1, two points of energy 2, two points of energy 3,...

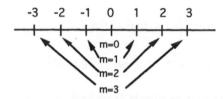

Figure 2.20.: Enumeration of points for the **Z** lattice: one point on the origin and two points for each non null index.

Since the $\mathbf{Z}^n$ cubic lattice is a Cartesian product of n one-dimensional lattices, its *Nu* function is a product of n v_Z functions multiplied by themselves.

$$v_{Z^n}(z) = \left(v_Z(z)\right)^n = \left(\frac{1+z}{1-z}\right)^n \qquad (2\text{-}37)$$

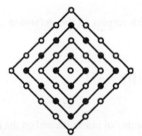

Figure 2.21.: Enumeration of points belonging to a pyramid of radius 4 for the $\mathbf{Z}^2$ lattice. We count the number of vectors on the pyramids of energy 0 (1 vector, white), of energy 1 (4 vectors, black), of energy 2 (8 vectors, white),..., and of energy 4 (16 vectors, white). The sum of these vectors is 41 (vectors contained within a pyramid of radius 4, energy = 4). See the table in Appendix B.

To obtain the function v_{D_n}, it is important to note that the lattice D_n is the set of points in $\mathbf{Z}^n$ whose L_1 norm is even. Hence v_{D_n} is the even part of v_{Z^n}, i.e.,

$$v_{D_n}(z) = \frac{1}{2}\left(v_{Z^n}(z) + v_{Z^n}(-z)\right) \qquad (2\text{-}38)$$

The definition of *Nu* functions for denser lattices is more complicated, and requires the use of binary codes theory.

Nu Series and A and B Constructions

Certain well-known lattices (Root lattices, Leech lattice) are built from algebraic codes. Several useful concepts regarding codes are recalled in order to introduce the definition of *Nu* series [82], [83]. Two definitions are then given, one for *Nu* series of lattices yielded by the *A* construction and another for *Nu* series of lattices yielded by the *B* construction.

The lattices used here are built from linear codes. Let $F_2 = \{0,1\}$ be a finite set of two elements. A *linear binary code* C is a vectorial subspace of F_2^n. Thus, C is a set of binary vectors of length n, and can be regarded as a subset of points in the hypercube unity. A code C is defined by the triplet (n, M, d):

n : number of coordinates of each basis vector (code length).
M : number of code words (code size or cardinal) denoted $|C|$.
d : minimum Hamming distance between two code words.

The code has at most 2^n vectors or code words C (all linear codes contain the element 0^n). A coset of C is defined by the code translated by the vector $r \in F_2^n$, $r + C = (r + C / C \in C)$.
The number of coordinates equal to 1 in a vector $\{C \in F_2^n\}$ is called the weight of C and is noted $W(C)$. Over F_2^n, the natural distance is the Hamming distance: $W(C) = |\{i / a_i = 1\}|$ where a_i is the ith component of the code word C.

Example:

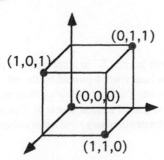

(0,1,1)
(1,0,1)
(0,0,0)
(1,1,0)

Three-dimensional parity check code $C(3,4,2)$. The code words are those whose sum is even { (0,0,0) , (0,1,1) , (1,0,1) , (1,1,0) }.

Let A_i be the number of code words of weight i in C. A_i represents the number of code words at the Hamming distance i from the origin (weight i).

$$A_i = |\{C \in C / W(C) = i\}| \quad \text{for} \quad i = 0,1, \dots, n$$

The list $\{A_i\}_{i=1,\dots,n}$ thus represents the weight distribution of the code C.

The *Hamming weight enumerator* of a code C is defined as the formal polynomial whose variables are x and y:

$$W_C(x,y) = \sum_{c \in C} x^{n-|c|} y^{|c|} = \sum_{i=0}^{n} A_i \, x^{n-i} \, y^i \tag{2-39}$$

This polynomial classifies the code words as a function of the number of non-null coordinates. The variable x counts the number of zeroes in a code word and y counts the number of ones. For the code $C(3,4,2)$, for example, the weight enumerator is the following:

$$W_{C(3,4,2)} = x^3 + 3xy^2 \text{ (one word of weight 0 and three words of weight 2).}$$

For greater detail on code words, the reader may consult references [27] and [55].

The definition of Nu functions is based on the weight enumerator (w.e.). The w.e. is associated with the lattice. We state hereafter the theorems for constructions A and B, and determine which of the two is better suited for the construction of the codebook.

A lattice can be built from a linear binary code. The lattice is associated with a code through a rule of construction. In the following, we recall constructions A and B, and show how to pass from the unit cube (which symbolizes the code) to an infinite lattice.

Construction A

Construction A is the simplest way to generate a lattice using a code C. Let C be a binary code (n,M,d). The following construction specifies a set of centers for packed spheres in $\mathbf{R}^n$.

$$A(C) := \left\{ Y = (y_1,\ldots,y_n) \in \mathbf{Z}^n \ / \ \exists c \in C, \ Y \equiv c \mod(2\mathbf{Z}^n) \right\} \tag{2-40}$$

Y is a center if and only if Y is congruent modulo 2 to a word in code C. In other words, a point Y, whose coordinates are integers, is a center if and only if the binary word formed by the bits of low weight of these coordinates expressed in base 2 belong to the code C (cf. the following diagram). Thus in order to generate the entire lattice, the motif of code C is placed at each center Y. The table of coordinates of a point $Y \in \mathbf{Z}^n$ is obtained by writing its coordinates in binary notation in columns in a table.

$$\begin{matrix} \boxed{0 \ 1 \ 0 \ 1 \ 0} & \textit{first row} \\ 0 \ 0 \ 1 \ 1 \ 0 \\ 0 \ 0 \ 0 \ 0 \ 1 \\ 0 \ 0 \ 0 \ 0 \ 0 \end{matrix}$$

$$\uparrow \ \uparrow \ \uparrow \ \uparrow \ \uparrow$$
$$Y = (0 \ 1 \ 2 \ 3 \ 4)$$

The assembly of the spheres yielded by construction A is a regular lattice if and only if the code C is linear.

The lattice D_n is obtained by the application of construction A to the even weight code $(n,2^{n-1},2)$ (parity check code). It is the set of vectors in the unit hypercube such that $W(C)$ is even, and is illustrated by the example presented earlier in 3 dimensions, $C(3,2^2,2)$.

The following theorem, demonstrated in [82], [83], enables us to determine the Nu functions of a construction A-type lattice, once the weight enumerator is known.

> **Theorem 1**: If C is a linear binary code whose weight enumerator is $W(x,y)$, and $A(C)$ the sublattice of Z^n whose vectors are congruent to C modulo 2, then the Nu function of the lattice is defined by:
>
> $$v_{A(C)}(z) = W_C\left(v_{2z}(z), v_{1+2z}(z)\right) = \frac{W_C\left(1+z^2, 2z\right)}{\left(1-z^2\right)^n} \qquad (2\text{-}41)$$

To illustrate this theorem, several examples of series are given below for the lattices Z^n, D_n and E_8.

Z^n LATTICE

If U_n is the universal code F_2^n, the weight enumerator (w.e.) is $W_{U_n}(x,y) = (x+y)^n$. U_n is the complete natural binary code; it is the complete set of points of the unit hypercube. If we substitute the value of this w.e. in formula (2-41), we obtain the Nu function for the lattice Z^n.

$$v_{Z^n}(z) = \frac{\left((1+z^2)+(2z)\right)^n}{\left(1-z^2\right)^n} = \frac{(1+z)^n}{(1-z)^n} \qquad (2\text{-}42)$$

This is the same result as that given in the preceding paragraph, found intuitively without using the concept of codes.

D_n LATTICE

It can be stated that $D_n = A(EW_n)$: application of construction A to the code EW_n where EW_n is the even weight linear binary code $(n,2^{n-1},2)$. It is the same code as the previous one, but with only even weight words. The weight enumerator is thus formed by the terms of even power in the w.e. of U_n. Actually, $D_n = 2Z^n + 2(n,2^{n-1},2)$.

$$W_{EW_n}(x,y) = \frac{1}{2}\left((x+y)^n + (x-y)^n\right) \qquad (2\text{-}43)$$

We thus obtain:

$$W_{EW_3}(x,y) = \frac{1}{2}\left((x+y)^3 + (x-y)^3\right) = x^3 + 3xy^2$$

$$W_{EW_4}(x,y) = \frac{1}{2}\left((x+y)^4 + (x-y)^4\right) = x^4 + 6x^2y^2 + y^4$$

If this w.e. is substituted into formula (2-41), the Nu function is the following:

$$V_{EW_n} = \frac{\frac{1}{2}\left[\left(\frac{1+z^2}{x} + \frac{2z}{y}\right)^n + \left(\frac{1+z^2}{x} - \frac{2z}{y}\right)^n\right]}{\left(1-z^2\right)^n}$$

Observing that $\dfrac{(1+z)^2}{1-z^2} = \dfrac{1+z}{1-z}$ and $\dfrac{(1-z)^2}{1-z^2} = \dfrac{1-z}{1+z}$, this expression simplifies to:

$$V_{D_n} = \frac{1}{2}\left(V_{Z^n}(z) + V_{Z^n}(-z)\right) \tag{2-44}$$

In particular, for the 4th dimension, the Nu function is:

$$V_{D_4} = \frac{1}{2}\left(V_{Z^4}(z) + V_{Z^4}(-z)\right)$$
$$= 1 + 32z^2 + 192z^4 + 608z^6 + \dots$$

E_8 LATTICE (*Gosset lattice*)

This lattice is obtained from the extended Hamming code H_8 of length 8 $(8,2^4,4)$ whose w.e. is $W_8(x,y) = x^8 + 14x^4y^4 + y^8$ (see [27] or [55]). Thus $E_8 = 2Z^8 + \left(8,2^4,4\right)$ and the Nu function is

$$V_{E_8}(z) = \frac{\left(1+z^2\right)^8 + 224z^4\left(1+z^2\right)^4 + 256z^8}{\left(1-z^2\right)^8} \tag{2-45}$$

Construction B

Construction B associates a lattice $B(\mathbf{C})$ with a code $\mathbf{C}$. It is defined by the following relationship:

$$B(\mathbf{C}) := \left\{Y = (y_1,\dots,y_n) \in Z^n \ / \ \exists \mathbf{c} \in \mathbf{C}, \ Y \equiv \mathbf{c} \mod(2Z^n) \ ; \sum_{i=0}^{n} y_i \equiv 0 \mod(4)\right\} \tag{2-46}$$

Let C be a binary code (n,M,d) containing words whose weight is even. In the B construction, a point Y is a center if and only if Y is congruent (modulo 2) to a word in code C and if $\sum_{i=1}^{n} y_i$ can be divided by 4.

Theorem 2: If C is a doubly even binary code whose weight enumerator is $W(x,y)$, then the Nu function is defined by

$$v_{B(C)}(z) = \frac{1}{2} W_C \left(\frac{1+z^2}{1-z^2}, \frac{2z}{1-z^2} \right) + \frac{1}{2}\left(\frac{1-z^2}{1+z^2}\right)^n \tag{2-47}$$

E_8 LATTICE

For this type of construction, the E_8 lattice is defined as the union of translated D_8 lattices: $E_8 = 2D_8 + r_i$ where r_i are the code words of the code $(8,2,8)$: $E_8 = (2D_8) \cup (2D_8 + (1,1,1,1,1,1,1,1))$. Thus, applied to the code $\mathscr{R}_8 = \{0^8, 1^8\}$, of w.e. $W = x^8 + y^8$, the above theorem leads to another Nu function for the E_8 lattice:

$$v_{E_8}(z) = \frac{1}{2} \frac{\left(1+z^2\right)^8 + 256z^8}{\left(1-z^2\right)^8} + \frac{1}{2}\frac{\left(1-z^2\right)^8}{\left(1+z^2\right)^8} \tag{2-48}$$

$$= 1 + 128z^4 + 2944z^8 + 1024z^{10} + O(z^{12})$$

The version of the E_8 lattice corresponding to this definition is half as dense as that obtained by applying construction A to the H_8 code. The Nu functions are identical if z is replaced by $z^{1/2}$.
Based on this result, the E_8 lattice will be used with construction B. The difference in the number of points can be interpreted as a rotation of the lattice about the construction axes. Indeed, although the lattice is the same, the axes are not positioned in the same manner depending on the construction. Hence the pyramids constructed from the axes do not contain the same number of points.

Λ_{16} LATTICE (*Barnes-Wall*)

This lattice is obtained by applying construction B to the 1st-order Reed-Muller code of length 16, with 32 code words $(16,2^5,8)$. The code words correspond to the 16 rows of the modified Sylvester-type Hadamard matrix H_{16} and their complements. The associated weight enumerator is $W_{16}(x,y) = x^{16} + 30x^8 y^8 + y^{16}$.
The following relationship enables the Nu function to be determined

$$v_{\Lambda_{16}}(z) = \frac{1}{2} \frac{W_{16}\left(1+z^2, 2z\right)}{\left(1-z^2\right)^{16}} + \frac{1}{2}\left(\frac{1-z^2}{1+z^2}\right)^{16} \tag{2-49}$$

$$= 1 + 512z^4 + 47872z^8 + \dots$$

The values of the coefficients N_m are given in Appendix B for the various lattices and different truncation radius values. Tables B.1. to B.10. were obtained by Taylor expansions of the generating functiond Θ and v. In these tables, the white rows show the number of points <u>on</u> a surface of radius m while the shaded rows give the number of points <u>within</u> a ball of radius m. These results were obtained using the Maple and Mathematica softwares.

In these examples, it is noted that for a constant radius, the pyramids are sparser than the spheres.

2.6.5.3. <u>Approximation of the Generating Series</u>

This paragraph decribes an approach to estimate the number of points contained in a codebook without directly using the generating series

$B_p(n,r)$ denotes the lattice points of energy less than or equal to r^p. This means that we consider a set of points inside a ball whose surface has a radius of r. The term "ball" is employed for all types of norm and corresponds to the quantization codebook. When $p=1$, the ball is a <u>pyramid</u>, and when $p=2$, it is a <u>sphere</u>.

The Gauss' enumeration principle indicates that the number of integer points (points whose coordinates are integers) within a convex body can be estimated by its volume. For a given Λ lattice, the number of points N in a ball is the ratio of the volume of this ball to the volume of the lattice's elementary Voronoi. This method is comparable to that used by Conway and Sloane to determine the density of a lattice (sphere packing), in which they use a sphere of unit radius.

$$N = \left| B_p(n,r) \cap \Lambda \right| \approx \frac{vol\big(B_p(n,r)\big)}{vol(\Lambda)}$$

(2-50)

The notation $\left| B_p(n,r) \cap \Lambda \right|$ indicates the cardinal of the intersection between the ball $B_p(n,r)$ and the lattice Λ used, i.e., the number of points of Λ contained in the ball. The numerator represents the volume of a ball as a function of its radius r, the dimension n, and the norm p. The denominator represents the volume of the lattice's elementary Voronoi region.

The <u>volume of the ball</u> is expressed by the relationship:

$$vol\big(B_p(n,r)\big) = 2^n r^n \, \frac{\Gamma\left(1+\dfrac{1}{p}\right)^n}{\Gamma\left(1+\dfrac{n}{p}\right)}$$

(2-51)

This formula is demonstrated in [62]. $\Gamma(.)$ is the Gamma function.

Two properties of the Gamma function, recalled hereafter, enable the usual formulas for the volume to be deduced, in the cases of $p=1$ and $p=2$:

$$\Gamma(n+1) = n! \quad \text{and} \quad \Gamma(n+1/2) = \frac{(2n)! \sqrt{\pi}}{2^{2n} n!} \quad \text{if } n \text{ is an integer}$$

In the case of the L_2 norm ($p=2$), formula (2-51) becomes

$$vol\big(B_2(n,r)\big) = 2^n r^n \frac{\Gamma\left(1+\dfrac{1}{2}\right)^n}{\Gamma\left(1+\dfrac{n}{2}\right)} = r^n \frac{\pi^{n/2}}{(n/2)!} = \frac{2^n \pi^{(n-1)/2}\big[(n-1)/2\big]!}{n!} \qquad (2\text{-}52)$$

If $n=2$, we find the area of a circle (πr^2) and if $n=3$, that of a sphere ($4\pi r^3/3$). In the case of the L_1 norm ($p=1$), formula (2-51) becomes

$$vol\big(B_1(n,r)\big) = \frac{2^n r^n}{n!} \qquad (2\text{-}53)$$

The denominator of formula (2-50) corresponds to the volume of the lattice's elementary Voronoi region, or the determinant of the lattice's Gram matrix. The volume is defined by the relationship $vol(\Lambda) = \left|\det GG^t\right|^{1/2}$ in which G is the generating matrix of the lattice (see paragraph 2.6.2.1.).

For example, for the D_4 lattice, the generating matrix is $G = \begin{bmatrix} 2 & 0 & 0 & 0 \\ 1 & 1 & 0 & 0 \\ 1 & 0 & 1 & 0 \\ 1 & 0 & 0 & 1 \end{bmatrix}$ and the determinant

$\det_{D_4} = \det GG^t = 4$, hence $vol(D_4) = 2$.

The ratio of volumes (2-50) yields the number of whole Voronoi cells in the ball. This approximation is based on the assumption that boundary problems are neglected. For the approximation to be valid, the intersection between the ball and the Voronoi regions at the boundary be small with respect to the total volume of the ball. This intersection is shown figure 2.22 by the shaded areas, which must be very small with respect to the circle's area for approximation (2-51) to be valid. This is all the more true when the radius m is large.

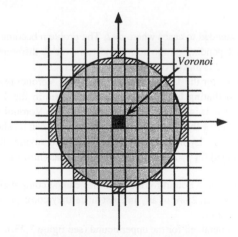

Figure 2.22.: Boundary problem: the Voronoi regions are superimposed on the ball

154

Upper and lower bounds can then be determined for the approximation. For the special case $p=2$, the upper bound N^+ on N is found by adding the covering radius $\mathcal{R}$ to the radius r of the ball (see figure 2.23). In this figure, the Voronoi cells cover the sphere $B_2(n, r-\mathcal{R})$.

Indeed, the Voronoi cells centered on the lattice points in the sphere $B_2(n, r)$ are included in the sphere $B_2(n, r+\mathcal{R})$. This leads to the following upper bound:

$$N^+ = \left| B_2(n, r+\mathcal{R}) \cap \Lambda \right| \approx \frac{vol\left(B_2(n, r+\mathcal{R})\right)}{vol(\Lambda)} \tag{2-54}$$

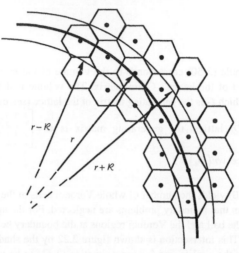

Figure 2.23.: Boundary problem for the approximation of the theta series. Truncation by spheres of radius $r - \mathcal{R}$, r, $r + \mathcal{R}$.

This concept is not easily extended to cases where $p \neq 2$. The problem becomes one of convexity, since the Voronoi regions differ depending on the norms and can overlap differently. In [82], a solution is given, in which an upper bound is placed on the covering radius $\mathcal{R}$.

Let $\mathcal{R}_2$ be the covering radius for the L_2 norm: the shape of the Voronoi region is determined by the L_2 norm (intersection of mediating hyperplanes) and the metric in the L_2 norm. Let $\mathcal{R}_p$ be the covering radius when the metric is in the L_p norm (the shape of the Voronoi region is still determined in the L_2 norm). Since the area of the Voronoi remains constant, it can be shown that the norms for p less than 2 are bounded by the L_2 norm [83]. This amounts to saying that the radii for $p \leq 2$ are bounded by the radius computed for $p=2$. $\mathcal{R}_p \leq \mathcal{R}_2$ for $0 < p \leq 2$, as shown in the following figure.

In cases where $0 < p \leq 2$, a maximum bound for the Nu series can be computed using $\mathcal{R}_2$. Moreover, in practice only values of p less than 2 are of any interest since they correspond to probability densities which are close to Laplacian laws.

The following relationship is obtained for the upper bound (see figure 2.23.):

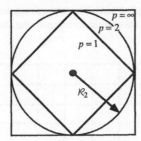

Figure 2.24.: The L_2 norm provides an upper bound for norms less than 2.

$$N^+ = \left| B_p(n, r+R_2) \cap \Lambda \right| = \frac{vol\left(B_p\left(n, r+R_2\right)\right)}{vol(\Lambda)} \geq N \tag{2-55}$$

Similar considerations enable the determination of a lower bound, which is given by:

$$N^- = \left| B_p(n, r-R_2) \cap \Lambda \right| = \frac{vol\left(B_p\left(n, r-R_2\right)\right)}{vol(\Lambda)} \leq N \tag{2-56}$$

Table 2.3. below summarizes the approximation formulas:

	N^-	N	N^+
$p=1$	$\dfrac{2^n\left(r-R_p\right)^n}{n!\,Vol(\Lambda)}$	$\dfrac{2^n r^n}{n!}$	$\dfrac{2^n\left(r+R_p\right)^n}{n!}$
$p=2$	$\dfrac{2^n\left(r-R_p\right)^n}{n!\,Vol(\Lambda)}$	$r^n \dfrac{\pi^{n/2}}{(n/2)!}$	$\left(r+R_p\right)^n \dfrac{\pi^{n/2}}{(n/2)!}$

Table 2.3.: Approximations of generating series.

Approximations for the *Nu* and *Theta* series based on the formulas in table 2.3. are shown in figure 2.25 for the D_4 lattice. The figure gives the exact series and the three approximations. It is easily verified that N^+ is indeed an upper bound and that N^- is a lower bound. Note also that, although N remains lower than the exact series curve, it cannot be considered an approximation by lower values. The higher and lower approximations for large m and fixed n are asymptotically equivalent to equation (2-50).

Although the curves seem to be divergent with respect to one another, they in fact converge toward the exact series. This convergence is easy to see graphically in a plot of the relative error.
Figure 2.26 shows the curves $(v-N)/v$ (curve (b)) and $\left(v-N^-\right)/v$ (curve (a)) as a function of the radius used. The relative error tends toward 0 as the radius increases. However, this error is too large

156

in the range of radii used for our applications. The approximations can therefore not be used instead of the exact series; for this reason, we are obliged to compute the *Nu* series.

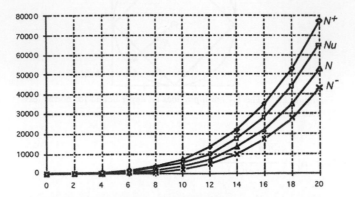

Figure 2.25.: Approximations of the theta series - Number of points within a pyramid versus the diameter of the pyramid.

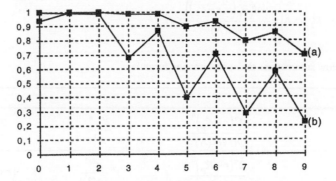

Figure 2.26.: Convergence of the series versus pyramid radius. (a) lower approximation - (b) medium approximation.

2.6.5.4. Comparison between Nu series and Theta series

For equal radii and sufficiently large size, pyramids contain far fewer points than spheres. In the case of cubic lattices, this fact can be attributed to the fact that the volume of an *n*-dimensional sphere of radius *m* is given by

$$\frac{\pi^{n/2}}{\Gamma\left(\frac{n}{2}+1\right)} m^{n},$$

while that of an *n*-dimensional pyramid is $\frac{2^{n}}{n!} m^{n}$, i.e., much smaller for a large *m*.

This result also shows that, for a constant radius, there are many more points in a sphere than in a pyramid as n increases.

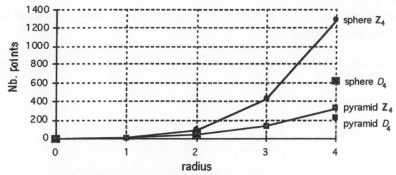

Figure 2.27.: Comparison between the number of points in pyramids versus spheres, for D_4 lattices.

2.6.6. Lattice Quantization Algorithm

In this paragraph, we present the complete quantization algorithm. We introduce the idea of <u>normalizing</u> the source by a scaling factor in order to adapt it to the codebook. Indeed, rather than using the scaling factor on the codebook, it is preferable to dilate or contract the source in order to use the fast quantization algorithms described in paragraph 2.6.3.

Lattice truncation and source normalization make it possible to control the bit rate and distortion.

2.6.6.1. Coding

Let m denote the radius of the ball defining the codebook. Quantization is performed in the following steps:

- Select the lattice according to the desired bit rate;
- Choose the size of the codebook (determine a lattice truncation radius m);
- Choose a maximum energy E_{max} for the source to be encoded (which may not necessarily be the source's maximum energy);
- Normalize the source using the projection factor γ, defined below for each source vector.

$$\gamma = \frac{E_{max}}{m} \text{ in the case of a } pyramid \text{ (Laplacian source)}$$

$$\gamma = \sqrt{\frac{E_{max}}{m^2}} \text{ in the case of a } sphere \text{ (Gaussian source).}$$

The projection factor enables quantization of the vectors belonging to the maximum energy source E_{max} by vectors in the peripheral zones of the codebook.

158

Modification of γ involves simply the value of the maximum energy to be encoded.
Source points whose energy exceeds E_{max} are projected separately onto the codebook surface (greatest energy) using different projection factors γ'.

- Quantize the vector by a lattice point using one of the algorithms described in paragraph 2.6.3.
- Index the quantized vectors.

<u>Reminder</u>: the cost of auxiliary data to be transmitted, i.e., the projection factors, is taken into account in the total bit rate.

2.6.6.2. <u>Decoding</u>

For each quantized vector:

- Decode its index;
- Find the point in the codebook associated with this index;
- Apply the scaling factor $1/\gamma$ for the vectors of energy less than m^p and γ' for vectors of energy equal to m^p ($p=1,2$).

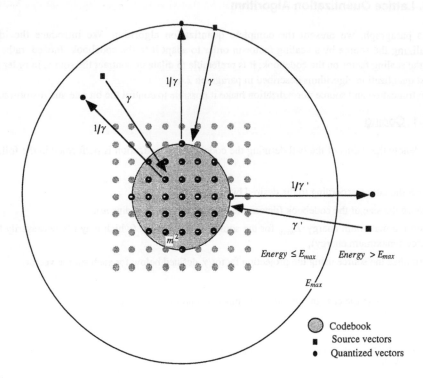

Figure 2.28.: Structure of a lattice vector quantizer with spherical truncation. The circles in the shaded area represent the vectors in the truncated lattice (codebook). The maximum energy of the vectors in the truncated lattice is m^2 and the maximum energy of the source vectors to be projected using γ is equal to E_{max}.

2.6.7. Labelling

Labelling is one of the last operations on the data before transmission or storage. This operation consists in assigning an index to each point in the codebook, in this case, to each point in the truncated lattice. This index is then converted into a binary word of length b_i bits, using an entropy coder (Huffman, arithmetic, etc...) or by construction of a product code (energy//[vector position] in the sphere or pyramid).

In the preceding paragraphs, emphasis was placed on the advantages of using lattices for vector quantization because the quantization algorithms are fast and simple. On the other hand, indexes must be computed; they are no longer addressed in a table as in the case of usual quantizers such as the LBG. There are several labelling or indexing algorithms [27], [52], [53]. Lamblin [52], for example, developed an algorithm based on the representation of a lattice as concentric spheres. The lattice points are arranged on the spheres as shown in figure 2.19. The index is the concatenation of the energy of the sphere and the order of the vector on this sphere. Another example is that of Conway and Sloane which consists in associating with each Λ lattice point located in a dilated Voronoi region $r\Lambda$ (r integer) the index vector of the coefficients on the basis defining the lattice.

In the product code (defined in paragraph 2.6.4.), indexing is quite simple. In this case, each lattice vector is identified by its coordinates in the lattice basis and is assigned to a code whose length corresponds to its energy level and its position on this level. The cost of indexing is then significantly reduced. The decoder must have the same transcoding table as the encoder (actually, it is able to reconstruct this table easily because the codebook is a lattice. The look-up table is no longer required to be transmitted (see [65]).

2.7. BIT ALLOCATION PROCEDURE

2.7.1. Problem

Let us define $D_T(R_T)$ the total distortion of a quantizer where R_T is the total bit rate we want to achieve. In a multiresolution scheme, we have the two following equations:

$$\begin{cases} D_T(R_T) = \sum_i a_i D_i(R_i) \\ \text{and } R_T = \sum_i a_i R_i \end{cases}$$

where a_i are weighting resolution parameters which depend on the multiresolution scheme we use[2] (see paragraph 1.7.). These parameters exist, because the implementation of the wavelet transform is not isometric. For example, in the classical dyadic case, $a_i=1/2^{2i}$.

---

[2] Note that, in the dyadic case, we must also consider a summation of the different orientation subimages (Horizontal - Vertical - Diagonal).

Each distortion $D_i(R_i)$ of each sub-image is defined as the distortion $D(R)$ given in paragraph 2.2.3, formula (2-7). The R_i correspond to the practical average bit rates allocated to each sub-image (sub-signal).

2.7.2. Classical Method

A classical method of optimization consists in minimizing the following functional using Lagrangian multipliers

$$J(R_i, \lambda) = D_T^*(R_T) + \lambda \left[\sum_i a_i R_i - R_T \right]$$

where $D_T^*(R_T) = \sum_i a_i D_i(R_i) B_i$ is the weighted distortion, and λ is a Lagrangian multiplier.
The assignment of the values B_i is based on the post-processing applied to the quantized signal (see paragraph 2.7.5.2. for image coding application).

The solution obtained here is analytical and is given by [4]

$$R_i^{opt} = \beta R_T + \frac{1}{2} \log_2 \left[\frac{s_i(p_i, 2, n_i) B_i}{\left(\prod_j \left(s_j(p_j, 2, n_j) B_j \right)^{1/4^j} \right)^{\beta}} \right]$$

where $s_i(p_i, 2, n_i) = A(n_i, 2) \left[\int_{\mathbb{R}} p_{X_i}(x_i)^{n_i/(2+n_i)} dx \right]^{(2+n_i)}$

β is a constant value which depends on the multiresolution analysis and is given in [4]. Furthermore, a problem remains in the values of R_i which could be negative.

2.7.3. Optimization with a Positivity Constraint

In order to avoid negative values, we introduce in the criterion a positivity constraint such that the bit allocation problem is formulated as

$$\underset{R_i}{\text{Min}}\, D_T^*(R_T)$$

subject to $\quad R_T = \sum_i a_i R_i$ $\qquad\qquad\qquad\qquad\qquad$ (2-57)

and $\quad 0 \le R_i \le R_{max}$

where R_{max} can be chosen as the maximum entropy of the sub-images. Then, the functionnal $J(R_i, \lambda)$ becomes

$$J(R_i,\lambda) = D_T^*(R_T) + \lambda\left[\sum_i a_i\,R_i - R_T\right] + \mu\left[\sum_i\left(\frac{|R_i|-R_i}{2}\right)^2 + \sum_i\left(\frac{|R_i-R_{max}|+R_i-R_{max}}{2}\right)^2\right]$$

where $\displaystyle\sum_i\left(\frac{|R_i|-R_i}{2}\right)^2$ ensures the positivity constraint

and $\displaystyle\sum_i\left(\frac{|R_i-R_{max}|+R_i-R_{max}}{2}\right)^2$ ensures the constraint $R_i \leq R_{max}$.

The choice for the value of μ, i.e. the positivity constraint in the functional, could result in a non-convex criterion. This is a typical ill-posed problem and the classical min max optimization methods may be caught in local minima.

2.7.4. A New Bit Allocation Scheme

However, a solution is given in [76], [77], [32 Chap.12] by the *augmented Lagrangian* method. The following functional must be minimized

$$J(R_i,\lambda) = D_T^*(R_T) + \lambda\left[\sum_i a_i R_i - R_T\right] + r\left[\sum_i a_i R_i - R_T\right]^2$$
$$+ \mu\left[\sum_i\left(\frac{|R_i|-R_i}{2}\right)^2 + \sum_i\left(\frac{|R_i-R_{max}|+R_i-R_{max}}{2}\right)^2\right]$$

(2-58)

where $\displaystyle\left(\sum_i a_i R_i - R_T\right)^2$ is the penalty function.

Here, we introduce a penalty function which ensures the existence of a saddle-point and the convergence of the algorithm, even for large values of μ. In fact, the trade-off between μ and r permits us to keep a convex criterion.

These values of μ and r are chosen by trial and error according to the global bit rate R_T we want to obtain.

Then, the minimization problem can be solved using a conjugate gradient method, and a solution is given by [63]

$$\underset{0\leq R_i \leq R_{max}}{\text{Min}}\ \underset{\lambda \in \mathbf{R}}{\text{Max}}\ \left[J(R_i,\lambda)\right]$$

Note that other kinds of augmented Lagrangian method, based on the same idea (use of the classical Lagrangian and penalty functions) has been proposed in the literature ([66] for example).

2.7.5. Application to a Multiresolution Image Coding Scheme

2.7.5.1. Principle

Multiresolution exploits the eye's masking effects and therefore enables us to refine the bit allocation according to the resolution level. Although a flat noise shape minimizes the MSE criterion, it is generally not optimal for subjective quality of images. To apply *noise shaping* across the subimages, we define a total weighted MSE distortion $D_T^*(R_T)$. The weights B_i included in this distortion measure are chosen according to the post-processing as described in paragraph 2.7.5.2.

The orthonormal property of the wavelet decomposition ensures an additive contribution of the quantization error (MSE) across the scales and directions. The normalization we choose for the wavelet coefficients and the low frequency coefficients introduces an increase of the distortion as a power of 4 in the dyadic case and as a power of 2 in the quincunx case (this is due to a non-isometric implementation of the wavelet transform). For example, in the dyadic case the parameters $a_i=1/2^{2i}$. Then, the total distortion and the global bit rate can be written :

$$
\begin{cases}
D_T^*(R_T) = \dfrac{1}{2^{2I}} D_I^{dc}\left(R_I^{dc}\right) B_I^{dc} + \displaystyle\sum_{i=1}^{I} \dfrac{1}{2^{2i}} \sum_{d=(H,V,D)} D_{i,d}\left(R_{i,d}\right) B_{i,d} \\[4mm]
\text{and } R_T = \dfrac{1}{2^{2I}} R_I^{dc} + \displaystyle\sum_{i=1}^{I} \dfrac{1}{2^{2i}} \sum_{d=(H,V,D)} R_{i,d}
\end{cases}
$$

where I stands for the lowest resolution (dc components) and $D_I^{dc}\left(R_I^{dc}\right)$ for the distortion of the lowest resolution with bit allocation R_I^{dc}.

For the computation of each theoretical distortion $D_i(R_i)$ we assume that the wavelet coefficient sub-images have Laplacian pdfs

$$
p_{X_i}(x_i) = \frac{1}{\sigma_i \sqrt{2}} e^{-\frac{\sqrt{2}}{\sigma_i}|x_i|}
\tag{2-59}
$$

Using formulas (2-8) and (2-59), each sub-image distortion can be easily computed and we can write :

$$
D_i(R_i) \le A(n_i,2) 2^{-2R_i} \left[2\sigma_i^2 \left(\frac{2+n_i}{n_i} \right)^{(2+n_i)} \right]
\tag{2-60}
$$

2.7.5.2. Choice for the Weighting Factors B_i

In general, the problem of finding an optimal bit allocation is formulated so as to minimize a distortion measure between the original image and the quantized image. However, the choice of the distortion measure is often motivated by a numerical evaluation of the quality of the quantized image (like Peak Signal to Noise Ratio or Peak SNR) and not necessarily by its subjective quality for example.

In this section, we propose a strategy to take into account both the post-processing applied on the quantized image : visualization, correlation … and the Modulation Transfer Function (MTF) of the imaging system. These quantities are introduced in the weighting factors B_i.

In fact, we assume that each MTF function is of the type $H(f_x, f_y)$ in the Fourier domain. Thus, we can write the weights B_i in the Fourier domain as a product of the considered MTFs since it corresponds to a convolution product in the spatial domain :

$$B(f_x, f_y) = \prod_{n=1}^{N} H_n(f_x, f_y)$$

if N is the number of MTFs we consider.

The MTF of the visual system is given in [46] and suggests that the human visual system is most sensitive to midfrequencies and least sensitive to high frequencies (see Figure 2.29.). The shape of the curve is similar to a band-pass filter, and Mannos and Sakrison [58] have shown that the frequency at which the peak occurs generally lies around 8 cycles/degree (8 c/d).

A curve fitting procedure [58] has yielded a formula for the frequency response of the visual system

$$H(f_x, f_y) = H(\rho) = A \left[\alpha + \left(\frac{\rho}{\rho_0} \right) \right] \exp\left[-\left(\frac{\rho}{\rho_0} \right)^{\beta} \right]$$

$$\rho = \sqrt{f_x^2 + f_y^2} \text{ cycles / degree}$$

(2-61)

where A, α, β, and ρ_0 are constants. For $\alpha = 0$ and $\beta = 1$, ρ_0 is the frequency at which the peak occurs. In an image coding application, the values $A=2.6$, $\alpha=0.0192$, $\rho_0=8.772$, and $\beta=1.1$ have been found useful by Mannos [58]. The sensitivity of the human eyes is plotted on figure 2.29.

To evaluate visual quality, the distance between the coded/decoded image and the observer is taken equal to $6H$, where H is the height of the image. Thus, spatial frequency of 56 cycles/degree corresponds to the Fourier frequency $f_e/2$.

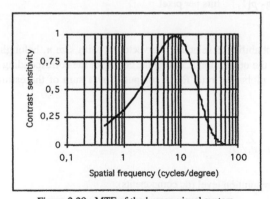

Figure 2.29.: MTF of the human visual system.

3. EXPERIMENTAL RESULTS

3.1. INTRODUCTION

3.1.1. Quality Criteria

The images used are sampled 512 by 512 pixel black and white images. The intensity of each pixel is coded on 8 bits which corresponds to 256 grey levels.

The numerical evaluation of the coder's performances is achieved by computing the peak signal-to-noise ratio (peak SNR) between the original image and the coded/decoded image. The formula of the peak SNR is given by

$$\text{Peak SNR}_{dB} = 10 \log_{10} \left(\frac{255^2}{\frac{1}{NM} \sum_{i=1}^{N} \sum_{j=1}^{M} \left(x_{i,j} - y_{i,j} \right)^2} \right) \tag{3-1}$$

where N and M correspond to the number of rows and columns of the image. x is the original pixel value and y its quantization.

Comparatively to the classical SNR, the peak SNR takes into account the peak value 255 of the original signal instead of its variance.

For each coded/decoded image we also give the corresponding bit rate R_T if an optimal entropy coding was performed (Shannon lower bound). As described in paragraphs 2.3 and 2.7.5.1, for each resolution level and edges orientation, we can introduce the average information of the codebook, called the entropy measure, computed by

$$H(Y) = -\frac{1}{n} \sum_{i=1}^{L} p(Y_i) \log_2 p(Y_i) \quad \text{bits per pixel} \tag{3-2}$$

where $p(Y_i)$ is the probability of selecting the vector Y_i with size n, belonging to the codebook at resolution 2^{-m} and for the orientation d, during the coding of the corresponding subimage. Then, as in paragraph 2.7.5.1, the global bit rate R_T is computed by a sum of the estimated entropy in each subimage as follows

$$R_T = \frac{1}{2^{2I}} R_I^{dc} + \sum_{i=1}^{I} \frac{1}{2^{2i}} \sum_{d=(H,V,D)} R_{i,d} \tag{3-3}$$

I stands for the lowest resolution level (dc components).

3.1.2. Choice of the Wavelet Bases

The *regularity* of the mother wavelet is important and appears to be closely related to the regularity of the signal to be processed. Since images are generally "smooth" to the eye, with the exception of occasional edges, it is appropriate to use regular and linear phase wavelets. Indeed, there is a trade-off between wavelet regularity and the visual effect on the coded/decoded image. The number of *vanishing moments* of the wavelet ψ, i.e., its oscillatory character, is also important. Actually, this number is related to the wavelet's regularity. In practice, a wavelet with N vanishing moments enables the cancellation of all wavelet coefficients of a polynomial signal whose degree is less than N. This result is quite significant for image coding applications because it enables high compression rates (many wavelet coefficients are zero or negligible).

According to the previous considerations and to the results obtained in [6], we use here, in all of our experimental results, the wavelets called "9-7" chosen among all the wavelet bases described in paragraph 1.6.5. In fact, the results presented in table 1.6 show that these wavelets offer the best trade-off between regularity and vanishing moments, and have the best spatial localization as defined formula (1-43).

3.2. LATTICE VECTOR QUANTIZER DESIGN

3.2.1. Choice of the Shaping

The knowledge of the source distribution is important for lattice vector quantization. In fact, it permits us to define the shape of the lattice truncation. Thus, by choosing a shape of truncation, which depends on the statistics of the source vectors, we adapt the quantizer to the data.

In practice, the wavelet coefficients lie on spheres, pyramids or ellipsoids depending on the source statistics and the multiresolution scheme we use. For example, quincunx multiresolution analysis provides in general wavelet coefficients with a multidimensional Laplacian statistical law [15]. The dyadic multiresolution scheme, instead, provides coefficients with a multidimensional Gaussian law or a product of monodimensional Gaussians with different standard deviations (ellipsoids) [16], [65]. Furthermore, all of this statistical laws are easily modeled by generalized Gaussian laws with parameters α less than or equal to 1.

Once the shape is chosen, we must find a truncation energy such that the number of vectors inside the codebook fits as well as possible the distribution of the source data in the space. If we increase this number of vectors, an hence the codebook size, we also increase the quality of the quantization results. However, the use of lattice vector quantizers permits us to increase the size without increasing the computational burden of the quantization algorithm. This is not the case for quantization methods based on classification like LBG. Thus, in our experiments, we use codebooks with sizes less than or equal to 65000 vectors. This maximal number is chosen according to the accuracy of calculation, which must be taken into account when using an entropy code, like Huffman or arithmetic code, for each codebook vector. However, the goal of this chapter is not to construct an entropy code. The problem of labelling a lattice vector and constructing the associated entropy code is solved by the use of product codes in [65].

This chapter focuses not on comparing different norms for the quantization, but rather on their use according to the multiresolution scheme and the source statistics. The interested reader should refer to [65] where a complete study comparing different norms (L_1, L_2, L_p) is provided.

3.2.2. Choice of the Lattice

In this section, we explain how to choose the lattice in each subimage in order to construct the multiresolution codebook.

The choice of the lattice results from a trade-off between the vector size and the desired bit rate obtained by the bit allocation procedure. The choice of the vector size is important because in each subimage we must be able to extract many vectors such that the statistical properties are still valid. For example, it is not judicious to extract vectors of size 16 pixels by 16 pixels in a subimage of size 32 pixels by 32 pixels.

Figure 3.1 shows a comparison between the SNRs provided by some lattices at low bit rates. We have compared the lattices Λ_{16}, E_8 and D_4 for a Laplacian source. We can see that lattice performance depends on the bit rate. For example, in our quantization scheme, the lattice Λ_{16} performs the best at a very low bit rate: this is due to the truncation of the lattice. In fact, if there were no overload noise, i.e., no truncation (huge codebook), then it would be possible to use a high dimension lattice like Λ_{16} for all of the different bit rates to achieve.

Thus, the choice of the lattice depends on the <u>resolution level</u> and the <u>bit rate allocation</u>.

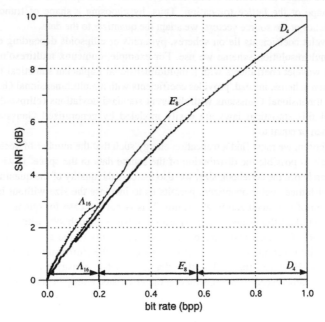

Figure 3.1.: Comparison between different lattices.

3.3. CODING WITH PSYCHOVISUAL CONSTRAINTS

For this experiment, we use a dyadic multiresolution scheme, and the image Lena (see figure 3.2) is decomposed into 4 resolution levels. According to the results of paragraph 3.2.1, the shapes of the lattice codebooks are taken spherical.

Here, two bit allocation schemes are shown. Table 3.1 corresponds to the bit allocation minimizing the usual MSE. Table 3.2 shows the bit allocation using visual criteria weights. To permit the algorithm to compute the bit rate values, we must choose the values of μ and r by trial and error.

The images on figures 3.3 and 3.4 are compressed images with a compression ratio of about 40:1, which corresponds to 0.2 bit per pixel (bpp). On figure 3.3 no weighting factors are used, while on figure 3.4 we use the visual weighting factors defined by formula (2-61). Although the peak SNRs are virtually the same, these images show that better coding results are obtained when using visual criteria. In fact, we can see that smooth edges (midfrequencies) are preserved in this case.

3.4. DYADIC OR QUINCUNX SCHEME FOR CODING ?

Here, we want to make a comparison between the dyadic and quincunx multiresolution schemes. According to paragraph 3.2.1, we quantize the wavelet coefficients resulting from the dyadic analysis by spherical codebooks, and from the quincunx analysis by pyramidal codebooks. For the comparison, we choose truncation energies such that the codebooks have the same size. The bit allocation is performed using a psychovisual criterion in both cases.

The coding results on the image Lena are showed on figures 3.4 and 3.5, and those obtained on Barbara are showed on figures 3.7 and 3.8.

For the Lena image the dyadic wavelet transform gives more ringing artefacts compared to the results obtained with a quincunx scheme. The separability of the filters, in the dyadic scheme, seems to cause a staircase on the edges.

This is not true for all kinds of images (see results on Barbara). In fact, quincunx bidimensional wavelets, more isotropic than separable dyadic wavelets, are well adapted to head and shoulder images. However, some images have horizontal and vertical structures like Barbara or satellite and robot vision images. For these kinds of images, a dyadic multiresolution scheme performs better than a quincunx one: see the zooms, figure 3.9.

Resolution levels	Orientation	Vector size n	Subimage variance	Weighting factors	Bit rate R (bpp)
1 (56 c/d)	Horizontal	8	2.764	1	0
	Vertical	8	3.842	1	0.170
	Diagonal	8	1.906	1	0
2 (28 c/d)	Horizontal	4	3.660	1	0.267
	Vertical	4	6.021	1	0.787
	Diagonal	4	3.138	1	0.175
3 (7 c/d)	Horizontal	4	4.522	1	0.443
	Vertical	4	8.141	1	1.340
	Diagonal	4	4.550	1	0.449
4 (1.75 c/d)	Horizontal	2	5.408	1	1.327
	Vertical	2	10.469	1	4.754
	Diagonal	2	5.103	1	1.199
4dc (0.44 c/d)	-	1	46.616	1	5

Table 3.1.: Bit allocation without weighting factors - $R_T \approx 0.2\,\text{bpp}$.

Resolution levels	Orientation	Vector size n	Subimage variance	Weighting factors	Bit rate R (bpp)
1 (56 c/d)	Horizontal	8	2.764	0.008	0
	Vertical	8	3.842	0.008	0
	Diagonal	8	1.906	0.008	0
2 (28 c/d)	Horizontal	4	3.660	0.232	0.277
	Vertical	4	6.021	0.232	0.680
	Diagonal	4	3.138	0.232	0.208
3 (7 c/d)	Horizontal	4	4.522	0.974	1.242
	Vertical	4	8.141	0.974	2.711
	Diagonal	4	4.550	0.974	1.252
4 (1.75 c/d)	Horizontal	2	5.408	0.480	1.622
	Vertical	2	10.469	0.480	4.794
	Diagonal	2	5.103	0.480	1.516
4dc (0.44 c/d)	-	1	46.616	1	5

Table 3.2.: Bit allocation with visual weighting factors - $R_T \approx 0.2\,\text{bpp}$.

Figure 3.2.: Original Lena image 512x512 pixels - $R_T = 8\,\text{bpp}$.

Figure 3.3.: Dyadic wavelet transform of Lena image and lattice vector quantization of wavelet coefficients. The allocation is performed without visual weighting factors.
$R_T \approx 0.2\,\text{bpp}$.

Figure 3.4.: Dyadic wavelet transform of Lena image and lattice vector quantization of wavelet coefficients. The allocation is performed with visual weighting factors defined by formula (2-61) and given in table 3.2. $R_T \approx 0.2\,\text{bpp}$.

172

Figure 3.5.: Quincunx wavelet transform of Lena image and lattice vector quantization of wavelet coefficients. The allocation is performed with visual weighting factors defined by formula (2-61). $R_T \approx 0.2\,\mathrm{bpp}$.

Figure 3.6.: Original Barbara image 512x512 pixels - $R_T = 8\,\mathrm{bpp}$.

Figure 3.7.: Dyadic wavelet transform of Barbara image and lattice vector quantization of wavelet coefficients. The allocation is performed with visual weighting factors defined by formula (2-61). $R_T \approx 0.4$ bpp. Peak SNR = 27.90 dB.

Figure 3.8.: Quincunx wavelet transform of Barbara image and lattice vector quantization of wavelet coefficients. The allocation is performed with visual weighting factors defined by formula (2-61). $R_T \approx 0.4\,\text{bpp}$. Peak SNR = 27.64 dB.

(a) Dyadic scheme

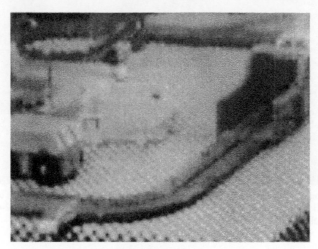

(b) Quincunx scheme

Figure 3.9.: Zoom on the coded/decoded Barbara image.
(a) Dyadic wavelet transform - (b) Quincunx wavelet transform.

APPENDIX B

NU AND THETA SERIES

In the case of a codebook comprised of a set of lattice points, the points contained within a sphere are identified by generatrix series Θ. The objective is to enumerate the vectors of energy m^2, i.e., those vectors located on the surface of the sphere of radius m.

The theta series is defined by

$$\Theta_\Lambda(q) = \sum_{Y\in\Lambda} q^{\|Y\|^2} = \sum_{m=0}^{+\infty} N_m q^{m^2} \quad \text{where} \quad q = e^{i\pi z} \tag{B-1}$$

If N_m denotes the number of vectors on the circle of radius m, the number of points contained *within* the sphere of radius m (surface of energy m^2) is then

$$N_T = \sum_{i=0}^{m} N_i$$

Similarly, N_T represents the coefficient of the term q^{m^2} in the modified series $\Theta_\Lambda(q)/(1-q)$. To enumerate lattice points contained within hyper-pyramids, we utilize the ν series, given by the following relationship

$$\nu_\Lambda(z) = \sum_{Y\in\Lambda} z^{\|Y\|_1} = \sum_{m=0}^{+\infty} \left[z^m\right]\left\{Y \in \Lambda / \|Y\|_1 = m\right\} \tag{B-2}$$

The similarities and differences with respect to the theta series below can be observed

$$\Theta_\Lambda(q) = \sum_{Y\in\Lambda} q^{\|Y\|^2} = \sum_{m=0}^{+\infty} \left[q^{m^2}\right]\left\{Y \in \Lambda / \|Y\|^2 = m^2\right\} \tag{B-3}$$

B.1. Z^2 Lattice

$$\Theta_{Z^2}(z) = \left[\theta_3(z)\right]^2 = 1 + 4z + 4z^2 + ... \tag{B-4}$$

d\u	0	1	2	3	4	5	6	7	8	9
0	1	4	4	0	4	8	0	0	4	4
	1	5	9	9	13	21	21	21	25	29
1	8	0	0	8	0	0	4	8	4	0
	37	37	37	45	45	45	49	57	61	61
2	8	0	0	0	0	12	8	0	0	8
	69	69	69	69	69	81	89	89	89	97

Table B.1.: Number of points in the Z^2 lattice <u>on</u> (white boxes) and <u>within</u> (grey boxes) a *sphere* of energy m^2 [$m = 10d+u$].

$$v_{Z^2}(z) = \left(\frac{1+z}{1-z}\right)^2 = 1 + 4z + 8z^2 + \dots \qquad \text{(B-5)}$$

d\u	0	1	2	3	4	5	6	7	8	9
0	1	4	8	12	16	20	24	28	32	36
	1	5	13	25	41	61	85	113	145	181
1	40	44	48	52	56	60	64	68	72	76
	221	265	313	365	421	481	545	613	685	761
2	80	84	88	92	96	100	104	108	112	116
	841	925	1013	1105	1201	1301	1405	1513	1625	1741

Table B.2.: Number of points in the Z^2 lattice <u>on</u> (white boxes) and <u>within</u> (grey boxes) a *pyramid* of energy m [$m = 10d+u$].

B.2. Z^4 Lattice

$$\Theta_{Z^4}(z) = \left[\theta_3(z)\right]^4 = 1 + 8z + 24z^2 + \dots \qquad \text{(B-6)}$$

d\u	0	1	2	3	4	5	6	7	8	9
0	1	8	24	32	24	48	96	64	24	104
	1	9	33	65	89	137	233	297	321	425
1	144	96	96	112	192	192	24	144	312	160
	569	665	761	873	1065	1257	1281	1425	1737	1897
2	144	256	288	192	96	248	336	320	192	240
	2041	2297	2585	2777	2873	3121	3457	3777	3969	4209

Table B.3.: Number of points in the Z^4 lattice <u>on</u> (white boxes) and <u>within</u> (grey boxes) a *sphere* of energy m^2 [$m = 10d+u$].

$$v_{Z^4}(z) = \left(\frac{1+z}{1-z}\right)^4 = 1 + 8z + 32z^2 + \dots \qquad \text{(B-7)}$$

d\u	0	1	2	3	4	5	6	7	8	9
0	1	8	32	88	192	360	608	952	1408	1992
	1	9	41	129	321	681	1289	2241	3649	5641
1	2720	3608	4672	5928	7392	9080	11008	13192	15648	18392
	8361	11969	16641	22569	29961	39041	50049	63241	78889	97281
2	21440	24808	28512	32568	36992	41800	47008	52632	58688	65192
	118721	143529	172041	204609	241601	283401	330409	383041	441729	506921

Table B.4.: Number of points in the Z^4 lattice <u>on</u> (white boxes) and <u>within</u> (grey boxes) a *pyramid* of energy m [$m = 10d+u$].

B.3. D_4 Lattice

$$\Theta_{D_4}(z) = \frac{1}{2}\left[\theta_3(z)^4 + \theta_4(z)^4\right] = 1 + 24z^2 + 24z^4 + \dots \tag{B-8}$$

d\u	0	2	4	6	8	10	12	14	16	18
0	1	24	24	96	24	144	96	192	24	312
	1	25	49	145	169	313	409	601	625	937
2	144	288	96	336	192	576	24	432	312	480
	1081	1369	1465	1801	1993	2569	2593	3025	3337	3817
4	144	768	288	576	96	744	336	960	192	720
	3961	4729	5017	5593	5689	6433	6769	7729	7921	8641
6	576	768	24	1152	432	1152	312	912	480	1344
	9217	9985	10009	11161	11593	12745	13057	13969	14449	15793

Table B.5.: Number of points in the D_4 lattice <u>on</u> (white boxes) and <u>within</u> (grey boxes) a *sphere* of energy m^2 [$m = 10d+u$]. Even values only.

$$\nu_{D_4} = \frac{1}{2}\left(\nu_{Z^4}(z) + \nu_{Z^4}(-z)\right) = 1 + 32z^2 + 192z^4 + 608z^6 + \dots \tag{B-9}$$

d/u	0	2	4	6	8
0	1	32	192	608	1408
	1	33	225	833	2241
1	2720	4672	7392	11008	15648
	4961	9633	17025	28033	43681
2	21440	28512	36992	47008	58688
	65121	93633	130625	177633	236321
3	72160	87552	104992	124608	146528
	308481	396033	501025	625633	772161

Table B.6.: Number of points in the D_4 lattice <u>on</u> (white boxes) and <u>within</u> (grey boxes) a *pyramid* of energy m [$m = 10d+u$]. Even values only.

B.4. E_8 Lattice

$$\Theta_{E_8}(z) = \frac{1}{2}\left[\theta_2(z)^8 + \theta_3(z)^8 + \theta_4(z)^8\right] = 1 + 240z^2 + 2160z^4 \ldots \tag{B-10}$$

d\u	0	2	4	6	8
0	1	240	2160	6720	17520
	1	241	2401	9121	26641
1	30240	60480	82560	140400	181680
	56881	117361	179921	320321	502001

Table B.7.: Number of points in the E_8 lattice <u>on</u> (white boxes) and <u>within</u> (grey boxes) a *sphere* of energy m^2 [$m = 10d+u$].

$$v_{E_8}(z) = \frac{1}{2}\frac{\left(1+z^2\right)^8 + 256z^8}{\left(1-z^2\right)^8} + \frac{1}{2}\frac{\left(1-z^2\right)^8}{\left(1+z^2\right)^8} = 1 + 128z^4 + 2944z^8 + 1024z^{10} + O(z^{12}) \tag{B-11}$$

d\u	0	1	4	6	8	10	12	14	16
0	1	0	128	0	2944	1024	31616	15360	199424
	1	1	129	129	3073	4097	35713	51073	250497

Table B.8.: Number of points in the E_8 lattice <u>on</u> (white boxes) and <u>within</u> (grey boxes) a *pyramid* of energy m [$m = 10d+u$].

B.5. Λ_{16} Lattice

$$\Theta_{\Lambda_{16}}(z) = \frac{1}{2}\left\{\theta_2(2z)^{16} + \theta_3(2z)^{16} + \theta_4(2z)^{16} + 30\theta_2(2z)^8\theta_3(2z)^8\right\}$$

$$= 1 + 4320z^4 + 61440z^6 + \ldots \tag{B-12}$$

d\u	0	2	4	6	8	10	12	14	16	18
0	1	0	4320	61440	522720	2211840	8960640	23224320	67154400	135168000
	1	1	4321	65761	588481	2800321	11760961	34985281	102139681	237307681

Table B.9.: Number of points in the Λ_{16} lattice <u>on</u> (white boxes) and <u>within</u> (grey boxes) a *sphere* of energy m^2 [$m = 10d+u$].

$$v_{\Lambda_{16}}(z) = \frac{1}{2}\frac{W_{16}\left(1+z^2, 2z\right)}{\left(1-z^2\right)^{16}} + \frac{1}{2}\left(\frac{1-z^2}{1+z^2}\right)^{16} = 1 + 512z^4 + 47872z^8 + \ldots \tag{B-13}$$

d\u	0	1	2	3	4	5	6	7	8	9	10
0	1	0	0	0	512	0	0	0	47872	0	92160
	1	1	1	1	513	513	513	513	48385	48385	14545

Table B.10.: Number of points in the Λ_{16} lattice <u>on</u> (white boxes) and <u>within</u> (grey boxes) a *pyramid* of energy m [$m = 10d+u$].

BIBLIOGRAPHY

V.R. Algazi, "Useful Approximation to Optimum Quantization", *IEEE Trans.commun.*, vol.COM-14, No.3, pp. 297-301 June 1966.

J.P. Adoul, M. Barth, "Nearest Neighbor Algorithm for Spherical Codes from the Leech Lattice", *IEEE Trans. on Inform. Theory*, vol. IT-34, pp. 1188-1202, 1988.

J. P. Adoul, "La quantification Vectorielle pour le Traitement de la Parole", *Conference organized by the GALF at ENST*: "La Quantification Vectorielle de Forme d'Onde: Approche Algébrique", pp 1-51, 14 February 1985. Paris.

M. Antonini, "Transformée en Ondelettes et Compression Numérique des Images", *ph-D Dissertation*, University of Nice-Sophia Antipolis, FRANCE, September, 1991.

M. Antonini, M. Barlaud, P. Mathieu, and I. Daubechies, "Image Coding Using Vector Quantization in the Wavelet Transform Domain", *IEEE ICASSP, Albuquerque USA*, pp. 2297-2300, April 1990.

M. Antonini, M. Barlaud, P. Mathieu, I. Daubechies, "Image Coding Using Wavelet Transform", *IEEE Trans. on Image Processing*, Vol.1, No.2, pp. 205-220, 1992.

M. Antonini, M. Barlaud, P. Mathieu, "Compression Numérique des Images par Quantification Vectorielle dans l'Espace des Transformée en Ondelettes", *Edition Masson*, 1992

M. Antonini, M. Barlaud, P. Mathieu, and J.C. Feauveau, "Multiscale Image Coding using the Kohonen Neural Network", *SPIE Visual Communication and Image Processing, Lausanne*, Switzerland, 1990.

M. Antonini, M. Barlaud, P. Mathieu, "Image Coding Using Lattice Vector Quantization of Wavelet Coefficients", *IEEE ICASSP, Toronto*, Canada, pp. 2273-2277, 14-17 May 1991.

M. Antonini, M. Barlaud, P. Mathieu, P. Solé, "Entropy Constrained Lattice Vector Quantization for Image Coding Using Wavelet Transform", *Workshop on computer vision and image processing for spaceborn applications, ESA, Noordwijk*, The Netherlands, 10-12 June 1991.

M. Antonini, M. Barlaud, T. Gaidon, "Adaptive Entropy Constrained Lattice Vector Quantization for Multiresolution Image Coding", *SPIE VCIP'92, Boston, Massachusetts*, USA, Vol. 1818, pp. 441-457, 18-20 November 1992.

M. Antonini, M. Barlaud, B. Rougé, C. Lambert-Nebout, "Weighted optimum bit allocation for multiresolution satellite image coding", $14^{\grave{e}me}$ *Colloque GRETSI, Juan les Pins*, France , pp. 455-458, September 1993.

[13] N. Baaziz, C. Labit, "Transformations pyramidales d'Images Numériques", IRISA, *preprint* n° 526, March 1990.

[14] M. Barlaud, P. Solé, M. Antonini, P. Mathieu, "A Pyramidal Scheme for Lattice Vector Quantization of Wavelet Transform Coefficients Applied to Image Coding", *IEEE ICASSP, San-Francisco, California*, USA, 23-26 March 1992.

[15] M. Barlaud, P. Solé, T. Gaidon, M. Antonini, P. Mathieu, "Pyramidal Lattice Vector Quantization for Multiscale Image Coding", To appear in April 1994 in *the IEEE Transactions on Image Processing*.

[16] M. Barlaud, P. Solé, JM. Moureaux, M. Antonini, P. Gauthier, "Elliptical Codebook for Lattice Vector Quantization", *IEEE ICASSP, Minneapolis, Minnesota*, USA, pp. 590-593, April 27-30, 1993.

[17] T. Berger, "Rate Distortion Theory", *Prentice Hall*, Inc., Englewood Cliffs, New Jersey, 1971

[18] A. Bijaoui, "Image et Information", *Edition Masson*, 1984.

[19] J. Bradley, T. Stockham, V.J. Mathews, "An Optimal Design Procedure for Subband Vector Quantizers", *submitted to IEEE Transactions on Communications*.

[20] P. Burt, E. Adelson, "The Laplacian Pyramid as a Compact Image Code", *IEEE Transactions on Communications*, Vol. 31 N°4, pp.532-540, April 1983.

[21] P. Charbonnier "Mise en Oeuvre d'une Chaîne de Compression d'Images dans un Environnement Convivial", *preprint I3S*, July 1991.

[22] P.A. Chou, T. Lookabaugh, and R.M. Gray , "Entropy Constrained Vector Quantization", *IEEE Trans. on ASSP*, Vol. 37, 1, pp. 31-42, 1989.

[23] J.H. Conway, N.J.A. Sloane, "Voronoi Region of Lattices, Second Moments of Polytopes, and Quantization", *IEEE Trans. on Information theory*, Vol. IT-28, No.2, March 1982.

[24] J.H. Conway, N.J.A. Sloane, "A Lower Bound on the Average Error of Vector Quantizers", *IEEE Trans. on Inform. Theory*, Vol. IT-31, No.1, pp.106-109, January 1985.

[25] J.H. Conway, N.J.A. Sloane "Fast Quantizing and Decoding Algorithms for Lattice Quantizers and Codes", *IEEE Trans. on Inform. Theory*, Vol. IT-28, No.2, pp. 227-232, March 1982.

[26] J.H. Conway, N.J.A. Sloane,"A Fast Encoding Method for Lattice Codes and Quantizers." IEEE Trans. on Inform. Theory, Vol. IT-29,6, pp. 820-824, 1983

[27] J.H. Conway, N.J.A. Sloane, "Sphere Packings, Lattices and Groups", *Springer Verlag*, 1988.

[28] D. Esteban, C. Galand, "Applications of Quadrature Mirror Filters to Split Band Voice Coding Systems", *ICASSP, Hartford*, USA, pp.191-195, May 1977.

[29] P. Filip, M.J. Ruf, "A Fixed Rate Product Pyramid Vector Quantization Using a Bayesian Model" *Proceedings of GLOBECOM'92, Orlando, Floride*, USA, Vol.1, pp. 240-244, 6-9 December 1992.

[30] T.R. Fischer,"A Pyramid Vector Quantizer" *IEEE Transactions on Information Theory*, Vol. IT-32, n°4, pp. 568-583, July 1986.

[31] T.R. Fischer, "Entropy Constrained Geometric Vector Quantization for Transform Image Coding", *IEEE ICASSP, Toronto*, Canada, pp. 2269-2271, 14-17 May 1991.

[32] R. Fletcher, "Practical Methods of Optimisation", *A Wiley - Interscience Publication*, 1987.

[33] J. Froment, "Traitement d'Images et Applications de la Transformée en Ondelettes", *ph-D Dissertation*, University of Paris IX Dauphine U.F.R Mathématiques de la décision, FRANCE, January 1990.

[34] T. Gaidon, M. Barlaud, P. Mathieu, "Utilisation d'Ondelettes Biorthogonales en Traitement des Images; Etude Théorique", *preprint I3S* n° 92-31, 1992

[35] T. Gaidon, M. Antonini, M. Barlaud, P. Mathieu, "Compression d'Images Fixes par Transformée en Ondelettes et Quantificateurs Vectoriels basés sur les Lattices", *preprint I3S*, n° 92-59, 1992

[36] T. Gaidon, "Quantification Vectorielle Algébrique et Ondelettes pour la compression de séquences d'images", *ph-D Dissertation*, University of Nice-Sophia Antipolis, FRANCE, December 1993.

[37] A.Gersho, "Asymptotically Optimal Block Quantization", *IEEE Trans. on Inform. Theory*, Vol. IT-25, No.4, pp. 373-380, July 1979.

[38] A. Gersho, "On the Structure of Vector Quantizers", *IEEE Trans. on Inform. Theory*, vol IT-28, No 2, March 1982.

[39] A. Gersho, R. M. Gray, "Vector Quantization and Signal Compression", Kluwer Academic Publishers, 1992.

[40] J.D. Gibson, K. Sayood, "Lattice Quantization", *Advances in Electronics and Electron Physics*, Vol. 72, pp 259-330, 1988.

[41] R.M. Gray, "Vector Quantization", *IEEE ASSP Mag.*, pp.4-29, April 1984.

[42] J. Hadamard, "Lecture on the Cauchy problem of linear partial differential equations". Yale University Press, 1923.

[43] Y. Ho, A. Gersho, " Variable-Rate Multi-Stage Vector Quantization for Image Coding " *IEEE ICASSP, New York*, USA, April 1988.

[44] D.A. Huffman, "A Method for the Construction of Minimum-Redundancy Codes", *Proceedings of the I.R.E.*, pp. 1098-1101, 1952.

[45] A. K. Jain, " Image data Compression: A Review", *Proceedings of the IEEE*, Vol 69, N°3, pp 349-389, March 1981

[46] A.K. Jain, "Fundamentals of Digital Image Processing", *Prentice Hall Information and System Sciences Series*, 1989.

[47] D.G. Jeong, J. D. Gibson," Lattice Vector Quantization for Image Coding", *IEEE ICASSP*, pp. 1743-1746, 1989.

[48] D.G. Jeong, J. D. Gibson," Image Coding with Uniform and Piecewise Uniform Quantizers, Preprint.

[49] G. Karlsson, M. Vetterli, " Three Dimensional Subband Coding of Video", *ICASSP*, pp. 1100-1103, 1988.

[50] G. Karlsson, M. Vetterli, " Extension of Finite Length Signals for Subband Coding", *Signal Processing*, Vol. 17, n° 2, pp. 161-168, June 1989.

[51] T. Kronander, "Some Aspect of Perception Based Image Coding", *ph-D Dissertation*, Linköping University, SWEDEN, 1989.

[52] C. Lamblin, "Quantification Vectorielle Algébrique Sphérique par le Réseau De Barnes-Wall: Application au Codage de la Parole", *ph-D Dissertation*, University of Sherbrooke, Quebec, Canada, March 1988.

[53] R. Laroia, N. Farvardin, " A Structured Fixed Rate Vector Quantizer Derived from a Variable Length Scalar Quantizer: Part I- Memoryless Sources, Part-II Vector Sources", *IEEE Trans. on Inform. Theory*, Vol. IT-39, No.3, pp. 851-876, May 1993.

[54] Y. Linde, A. Buzo, R. M. Gray, "An Algorithm for Vector Quantizer Design", *IEEE Trans. on Comm.*, Vol. COM-28, No.1, pp. 84-95, January 1980.

[55] J. H. van Lint, "Introduction to Coding Theory", *Springer Verlag*, 1991.

[56] S.P. Lloyd, "Least Squares Quantization in PCM", *IEEE Trans. on Information Theory*, Vol. IT-28, pp. 129-137, March 1982.

[57] B. Macq, "Weighted Optimum Bit Allocations to Orthogonal Transforms for Picture Coding", *IEEE Jour. Sel. Areas in Communications*, 10-5, pp. 525-536, June 1992.

[58] J.L. Mannos and D.J. Sakrison, "The effects of a visual Fidelity Criterion on the Encoding of Images", *IEEE Transaction on Information Theory*, IT-20, No. 4, July 1974.

[59] N. Maschio, "Contribution à la Compression d'Images Numériques par Codage Prédictif et Transformée en Cosinus Discrête avec Utilisation de Codes Arithmétiques", *ph-D Dissertation*, University of Nice Sophia-Antipolis, France, July 1989.

[60] P. Mathieu, M. Barlaud, M. Antonini, "Compression d'Images par Transformée en Ondelette", *12ième colloque GRETSI, Juan les Pins*, FRANCE, June 12-16, 1989.

[61] P. Mathieu, M. Barlaud, M. Antonini, "Compression d'Image par Transformée en Ondelette et Quantification Vectorielle", *Traitement du Signal*, vol. 7, n°2, pp.101-115, 1990.

[62] J.E. Mazo, A.M. Odlyzko, "Lattice Points in High-Dimensional Spheres", Mh. Math. 110 pp. 47-61, 1990.

[63] M. Minoux, "Programmation Mathématique - Théorie et Algorithmes- Tome 1", *CNET-ENST, Collection Technique et Scientifique des Télécommunications*, éditions DUNOD, 1983.

[64] P. Monet, C. Labit, "Codebook Replenishment in Classified Pruned Tree-Structured Vector Quantization of Image Sequence", *IEEE ICASSP, Albuquerque*, USA, pp.2285-2288, April 1990.

[65] J.M. Moureaux, M. Antonini, M. Barlaud, "New Lattice Vector Quantization Design for Image Coding", *submitted to Image Processing*, July 1994.

[66] H. Nakayama, H. Sayama, Y. Sawaragi, "A Generalized Lagrangian Function and Multiplier Method", *Journal Optimization Theory and Applications*, Vol. 17, No. 3/4, pp. 211-227, 1975.

[67] N.M. Nasrabadi, R.A. King, "Image Coding Using Vector Quantization: A Review", *IEEE Trans. on Comm*, Vol 36, No 8, August 1988.

[68] E. Nguyen, C. Labit, "Définition Quantitative des Matrices de Pondération Psychovisuelles pour la Quantification Adaptée en Codage Sous-Bandes d'Images", *14ème Colloque GRETSI*, *Juan les Pins*, France , pp. 419-422, September 1993.

[69] M.G. Perkins, T. Lookabaugh, "A Psychophysically Justified Bit Allocation Algorithm for Subband Image Coding Systems", *IEEE ICASSP*, Glasgow, Scotland, pp. 1815-1818, 23-26 May 1989.

[70] W.K. Pratt, "Digital Image Processing", *J.Wiley, New-York*, 1978.

[71] K. Ramchandran, M. Vetterli, "Best Wavelet Packet Bases in a Rate-Distortion Sense", *IEEE Transactions on Image Processing*, Vol.2, N°2, pp. 160-175, April 1993.

[72] R. Rao, W. Pearlman, "Multirate Vector Quantization of Image Pyramids", *ICASSP*, pp 2257-60, 1991

[73] O. Rioul, "Ondelettes Régulières : Application à la Compression d'Images Fixes", *ph-D Dissertation, Ecole Nationale Supérieure des Télécommunications*, FRANCE, March 1993.

[74] E. Riskin, E. M. Daly, R. M. Gray, "Pruned Tree-Structured Vector Quantization in Image Coding " *IEEE ICASSP, Glasgow*, SCOTLAND, pp. 1735-1738, May 1989.

[75] Rissanen, K.M. Mohiuddin, "A Multiplication-Free Multialphabet Arithmetic Code", *IEEE Trans. on Comm.*, Vol.37, No.2, pp. 93-98, February. 1989.

[76] R.T. Rockafellar, "A Dual Approach to Solving Nonlinear Programming Problems by Unconstrained optimization", *Mathematical Programming 5*, pp. 354-373, 1973.

[77] R.T. Rockafellar, "Augmented Lagrange Multiplier Functions and duality in Nonconvex Programming", *S.I.A.M. Journal Control 12*, pp. 268-285, 1974.

[78] T. Senoo, B. Girod ,'"Vector Quantization for Entropy Coding of Image Subbands", IEEE Trans. on Image Processing No 4, pp. 526-533, October 1992.

[79] C.E. Shannon, "A mathematical theory of communications", *Bell systems Technical Journal 27*, pp. 379-423,623-656, 1948.

[80] C.E. Shannon, "Coding Theorems for a Discrete Source with a Fidelity Criterion", *IRE Nat. Conv., Pt. 4*, pp. 142-163, 1959.

[81] N.J.A. Sloane, letter to Patrick Solé, 1th August 1991.

[82] P. Solé, "Counting Lattice Points in Pyramids", Actes de congres Séries Formelles et Combinatoire Algébrique (UQUAM Research Report), Montréal, June 1992. *To appear in Disc. Math.*

[83] P. Solé, "Generalized Theta Functions for Lattice Vector Quantization", *Dimacs congress on coding and quantization*, Rutgers, USA, 1992.

[84] Tanabe, Favardin, "Subband Image Coding Using Entropy - Coded Quantization over Noisy Channels", *IEEE Journal on Selected Areas in Communications*, Vol. 10, N° 5, June 1992.

186

[85]   H. Tseng, T.R. Fischer, "Transform and Hybrid transform /DPCM Coding of images using pyramid vector quantization", IEEE COM-35,pp. 79-86, 1987.

[86]   M. Vetterli, "Multidimensional Subband Coding: Some Theory and Algorithms", *Signal Processing* 6, 1984.

[87]   P.H. Westerink, D.E. Boekee, J. Biemond, J.W. Woods, "Subband Coding of Image Using Vector Quantization", *IEEE Trans. on Comm.*, Vol. 36, pp. 713-719, 1988.

[88]   J. Woods, O. Neil, "Subband Coding of Images", *IEEE Trans on ASSP*, Vol 34 No 5, pp. 1278-1288, October 1986.

[89]   P. Zador, "Asymptotic Quantization Error of Continuous Signals and their Quantization Dimension", *IEEE Trans. on Information Theory*, vol.IT-28 , 1982.

4. CONCLUSION

The main study presented here is the association of the wavelet transform and lattice vector quantization for image coding applications.

Today, the wavelet transform is a well-known method for the signal processing community. In the first section of this chapter we simply recalled the main properties of this transform with a complete bibliography in order to help the interested reader. We also presented our work with Ingrid Daubechies to design wavelet bases adapted to image coding. The best wavelet basis resulting from this study is called, in this chapter, "9-7". This compactly supported wavelet basis offers the best trade-off between regularity and vanishing moments, the associated filters have short length and linear phase. According to our knowledge and our coding results, today, this basis is one of the best for image coding applications.

Our main contribution was to develop lattice quantizers adapted to the statistics of the source which is constituted by wavelet coefficients. In the second paragraph of this chapter, we explained what a lattice is and how it is used to construct a vector quantizer. Furthermore, we proposed a new pyramidal shape for the truncation of the lattice adapted to Laplacian statistics. Indeed, previous work on lattices considered Gaussian sources only. Today, pyramidal codebooks are well adapted to wavelet coefficients, like quincunx wavelet coefficients. For this kind of shaping, we proposed a method to compute the number of lattice points lying on pyramids. This method is based on the works of Conway and Sloane for Gaussian distribution and extends the method of Fischer, developed on Laplacian sources for the Z^n lattice, to other lattices like the D_4, E_8 and Λ_{16} lattices.

Some of our coding results were presented in section 3. These results show that a compression ratio of about 40:1 can easily be obtained. For the future, the challenge is to obtain compression ratios more than 100:1 with similar image quality.

Now, in order to improve our coding results, we must improve both the choice of the wavelet bases and the quantizer design. The main problem is to define a wavelet basis efficient for the compression and adapted to the image. This wavelet basis must be optimized according to criteria such as the number of vanishing moments, regularity, spatial variance of the wavelets or criteria based on the shape of the wavelets. Not only the wavelet basis, but also the quantizer design must be improved. In fact, the question is: which quantizer, adapted to the statistics of the source, could provide minimal distortion with the lowest possible entropy ? However, the quality of a quantizer is not only its capability to provide low distortion. In fact, a quantizer could also take into account the pre and post-processing performed on the coded/decoded image. In other words, the design of the quantizers will be different if we want to visualize the coded/decoded images or if we want to do stereo-matching on it, for example.

188

ACKNOWLEDGMENTS

The authors wish to thank *Pr. Ingrid Daubechies* from AT&T Bell laboratories for her help in the works on wavelet transform, and *Dr. Patrick Solé* from I3S/CNRS laboratory for his contribution to the works on lattices. The authors wish to thank also *Dr. Laure Blanc-Féraud* from I3S/CNRS laboratory for helpful discussions.

Chapter 4

A Region-Based Discrete Wavelet Transform for Image Coding

H. J. Barnard, J. H. Weber, and J. Biemond

Information Theory Group
Department of Electrical Engineering
Delft University of Technology
P.O. Box 5031, 2600 GA Delft
The Netherlands

Abstract

Region-based image coding methods divide the image into regions and code the segmentation information and the texture separately. We introduce a region-based discrete wavelet transform (RBDWT) to code the texture. Its advantages in respect to other texture coding methods are its low computational complexity and its ability to code the remaining edges inside the regions. First, we show the principles of the RBDWT and explain the details for implementation. Then, the performance of the new method is evaluated for two test images. The subjective and objective coding performances of the RBDWT are compared to the standard discrete wavelet transform (DWT). It turns out that the signal-to-noise ratio is slightly lower for the new method, but the subjective visual quality is better. Therefore, the new image coding technique is at least competitive to the standard DWT. Finally, suggestions are given for further research in this new direction.

1. INTRODUCTION

1.1. The discrete wavelet transform

The discrete wavelet transform (DWT) is used in many areas of signal processing. Its application to image coding was described in [11] for the first time. In [1] finite length biorthogonal wavelet filters were used, which are well suited to image coding. The application of the wavelet theory to finite length discrete signals is equivalent to the older subband decomposition, which was introduced for image coding in [17] and investigated thoroughly in [16]. The reason for dividing the image into subbands is that the available bit rate can be distributed over the subbands according to their relative importance for the human visual system (HVS). The main difference between the two theories resides in the filters that are used. In [2] the effect of the filter choice on the coding performance is evaluated. It turns out that the differences are not very large, but that the biorthogonal filter pairs are best.

1.2. Region-based coding

Region-based, or contour-texture, coding techniques have a different approach to coding images than the transform coding methods as the DWT. The idea of region-based coding is to divide the image into regions, which correspond, as much as possible, to the objects in the image. This type of coding method has been extensively discussed in [9]. With region-based coding, the edges in an image are preserved better than with transform coding.

For the division into regions, the segmentation, several algorithms have been proposed during the last decades, for instance, region-growing and split-and-merge methods [10]. The segmentation information can be coded with a chain-code [7], which under certain assumptions may only cost 1.27 bits per contour node [5].

Most often, the texture information is transmitted by the coefficients of a 2-D polynomial that is fitted to the segment data. Generally, this polynomial is chosen to be of degree 0, 1, 2, or 3, depending on the bit rate and the region properties, such as variance and size. Another method for texture coding is the use of generalized orthogonal transforms, introduced in [8]. Here, the basis images of the transform are calculated by orthogonalizing standard basis images with respect to the shape of the region. The weights of the basis images are transmitted after they have been quantized to get at the desired bit rate. In both cases, the available bit rate is distributed over the regions according to their relative importance.

We end this section by remarking that the coding performance of any region-based method relies on the quality of the segmentation. Therefore, the segmentation process should always be optimized with regard to the texture coding method.

1.3. Principle of the region-based discrete wavelet transform

The principle of a region-based discrete wavelet transform can be viewed as a combination of the two coding principles of the previous sections. The DWT decomposes the image into subbands, each of which gets from the bit allocation algorithm a number of bits corresponding to the importance of the subband. In contrast, region-based coding methods are based on the division of the image into regions and each region is coded with an amount of bits that reflects its relative importance in respect to the other regions. Figure 1 illustrates that a region-based coding scheme using the DWT combines both the division into regions and the division into subbands. Now, the bit allocation algorithm is able to distribute bits over all *subband regions*, such that important features in the spatial as well as the frequency domain can be represented accurately.

In other words, we propose here a new method for coding the texture in a region-based coding scheme: the DWT. The standard DWT as described in literature cannot be used to code the texture. The reason for this is that a d-level decomposition requires the regions to consist of square blocks of size $2^d \times 2^d$. In general, such regions will not correspond to the shape of the objects in a scene. However, the efficient signal extension method described in [3] and summarized in Section 2.2 enables us to apply a DWT on arbitrarily sized regions. To distinguish this implementation of the DWT from the standard DWT, we use the name region-based DWT (RBDWT).

The main difference of the RBDWT in comparison to the texture coding methods mentioned in Section 1.2 is that the RBDWT requires relatively large regions. The polynomial fitting method needs small regions because polynomials cannot take care of variations inside a segment. For the method of generalized orthogonal transforms, the computational burden is the most important reason to keep the regions small-sized. In contrast, the RBDWT is perfectly able to represent remaining edges and discontinuities inside regions and the computational complexity is very low. When the regions are too small, the RBDWT is of no use as the transmitted coefficients will only consist of the mean values of the segments. In that case, the advantage of the RBDWT in that it decomposes the signal into different subbands is not used. Therefore, the regions can be and should be relatively large for the RBDWT, yielding a lower coding overhead for the transmission of the segmentation.

1.4. Scope and organisation

The subject of this chapter is to explain in detail the implementation of the RBDWT and compare the coding performance of this new image coding technique with the standard DWT. As the method is relatively recently developed, a complete evaluation cannot be presented yet, but from the primary results we conclude that the method is worth further investigation.

The remainder of this chapter is divided into four parts. Firstly, we present in Section 2 the details of the implementation of the RBDWT. In this part we explain how a signal of arbitrary length can be decomposed into subbands without an increase in data. This type of decomposition forms the basis of the RBDWT. Secondly, in Section 3 the coding scheme is described with which we performed our experiments. In the next part, Section 4, we discuss the advantages and disadvantages to be expected for the RBDWT, in comparison to the standard DWT. We also show the coding performance both objectively and subjectively evaluated, and we discuss the application of the RBDWT to video coding. Finally, we summarize the conclusions and give recommendations for further research in Section 5.

2. THE IMPLEMENTATION OF THE RBDWT

In the following, we describe in detail the practical implementation of the RBDWT. We assume that the segmentation into regions has already been performed by a proper split-and-merge method, taking into account that the texture is coded by the RBDWT, that is, a division into regions which are not too small (Section 1.3). We start in Section 2.1 by explaining how to decompose the regions of the original image, the *segmentation mask*, into four subbands, while preserving the possibility of perfect reconstruction. In Section 2.2 we describe how the filtering inside regions takes place.

2.1. Decomposition of the segmentation mask

The standard (dyadic) 2-D discrete wavelet transform for images decomposes an image into four subbands (called LL, LH, HL, and HH), by applying a two-channel splitting both on the

(a)

(b)

Figure 1   Schematic illustration of the RBDWT being a combination of the DWT and a region-based coding method. (a) Original image, split up into: (b) subbands (as used by DWT); (c) regions (as used in region-based coding); (d) subbands and regions (as used in RBDWT).

(c)

(d)

Figure 1 (Continued).

rows and on the columns (Figure 2), and this decomposition is repeated recursively on the LL subband.

Even when the image is divided into regions, we still wish to be able to decompose the image into four subbands, where each subband is subdivided into regions as well. Four 'child' regions of decomposition level 1 correspond to one 'parent' region of the original image. The division of the subbands into regions has the restriction that each parent region in the original image must be perfectly reconstructible from the four corresponding child regions in the subbands. This means that the total number of subband pixels in the four child regions must be at least equal to the size of the parent region, otherwise some information will be lost. Since the total number of subband pixels remains the same, equality is required.

The above requirement for the segmentation of the subbands can be achieved with the method as demonstrated in Figure 3. The segmentation mask of the original image is subdivided into 2x2 blocks and for each block the upper-left part is put into the LL band. In the same way, the upper-right part of each block falls in the LH band, the lower-left into the HL band, and finally, the lower-right part is copied into the HH band. This procedure guarantees that the total number of pixels in one region remains the same. The procedure can be viewed as an algorithm to downsample in four ways a segmentation by a factor 2 per dimension. We see in every subband in Figure 3c a kind of jig-saw puzzle in which the pieces exactly fill up the rectangular shape of the subband. Note that the two congruent regions in the center are divided over the subbands in a different way, because of the different parity of the pixel coordinates. Of course, the decomposition can be repeated on any of the subbands, although we apply it only on the LL subband, as with the wavelet decomposition.

2.2. Filtering and downsampling with the RBDWT

Conventional methods of signal extension that are used in the filtering of finite length signals depend on the signal being of even length N. Then the two subband signals are of length $N/2$. However, for the RBDWT the regions may have lines and columns that are of odd

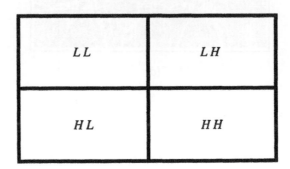

Figure 2   A splitting into four subbands. The LH subband contains vertical low-pass and horizontal high-pass information, and so on.

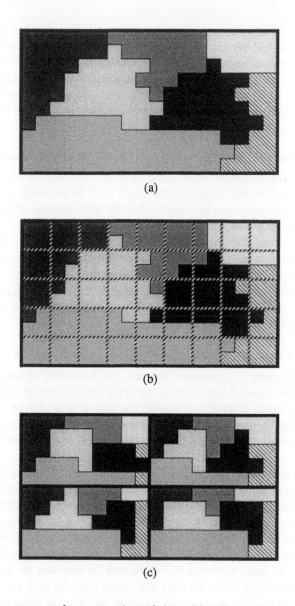

(a)

(b)

(c)

Figure 3   The process of segmentation of the subbands; (a) The segmentation of the
original image; (b) The subdivision into 2x2 blocks; (c) The segmentation of the subbands.

length. Since we do not allow an increase in data we have to decompose the odd length signal in two subband signals that differ in length by 1. In order to achieve this, we apply the efficient signal extension method as explained in [3], which preserves the perfect reconstruction (PR) property. Firstly, we define the general condition on signal extension that must be fulfilled in order to preserve the PR property [12]. Then, we show in two examples how signal extension should be performed for even and odd length signals. Finally, we describe the filtering of segment lines.

2.2.1. General condition on signal extension for enabling perfect reconstruction

Filtering of finite length signals causes problems at the boundaries of the signal because the filter crosses the signal border. Therefore, the signal must be extended at the boundary, for which several methods exist some of which enable PR. The criterion of whether a boundary extension allows PR can be stated as follows.

Suppose, the signal of length N is extended to infinity at both boundaries, filtered, and downsampled. If all the coefficients of both infinite length subbands can be determined from a subset of N/2 samples in each band then the extension enables PR.

Somewhat more restrictively, we can say that only $N/2$ different samples may appear in each infinite length subband. The combination of linear phase PR filters and the symmetric extension method is most suitable for image coding.

2.2.2. Example of symmetric signal extension for even length signals

In Figure 4 we illustrate in a schematic diagram how symmetric extension for odd length symmetric filters takes place. It shows a two-channel analysis/synthesis scheme where on the left-hand side the low-pass channel is displayed. The transmission channel comprises also the coding and decoding. The four large blocks contain a pictorial description of the filtering. To demonstrate the extension we use a biorthogonal filter pair of lengths 9 and 7 that enable perfect reconstruction.

In both analysis blocks, the top row shows the original signal of length $N = 8$ in the dark rectangle. Each little square denotes a sample the value of which is represented by a lower case letter. Outside the dark rectangle, the signal is extended symmetrically, which is visualized by using the same lower case letters when the same signal value appears. The $N/2 = 4$ rows in the middle represent the four filter positions. The filter coefficients are written in upper case letters, and the symmetry of the filters is symbolized in the way the letters are repeated. For each filter position, the inner product is calculated with the filter and the part of the extended signal in the top row directly above the filter. The values from the $N/2$ inner products are put into the bottom row in the dark rectangle. For example, the 'k' is the result of the inner product of the filter H in its first filter position with the part of the signal denoted by the symbols 'e', 'd', ..., 'e'. The downsampling is illustrated by the empty squares. In the bottom row, it is also shown which values would have appeared outside the dark rectangle if the convolution would have been over more than $N/2$ filter positions. We observe that no new values appear, which means that perfect reconstruction is possible: all information in the subbands is contained in the samples in the dark rectangles.

In the synthesis blocks, the subband signal is extended and upsampled as is demonstrated in the top row. All N filter positions are needed for the reconstruction (only 3 are shown). Note, that the type of symmetry depends on the channel. For one channel, the left-hand side is extended with odd symmetry and the right-hand side with even symmetry. For the other channel the reverse holds. Thus, the synthesis extension is different from the analysis extension [12]. The delay of the total system is 0, which can be checked from the bottom rows.

2.2.3. Example of symmetric signal extension for odd length signals

We now show how a signal of odd length N can be represented by a number of subband samples equal to N. We do this by implementing an efficient manner of signal extension for symmetric odd length filters (Figure 5).

Our signal is length $N = 7$ and is represented by the dark rectangle enclosing the samples denoted by 'a' through 'g'. The new extension method consists of adding one specific sample 'h' and then extending the signal symmetrically as before. The value of the sample 'h' in the extension is chosen such that one of the samples in the subbands becomes a fixed value. This fixed value is signal-independent and therefore contains no information about the signal. As both the transmitter and the receiver know this fixed value beforehand, it does not need to be transmitted and so the number of subband samples equals the length of the original signal.

In our example, the fixed sample corresponds to the rightmost sample of the high-pass subband in Figure 4, which is denoted by 'w'. Its value is the result of the filtering with the high-pass filter in the last of the four filter positions and forms a linear equation with one unknown variable 'h'. In Figure 5 the value of 'w' has been fixed to 0 and 'h' is uniquely determined.

On the receiver side, no additional information is needed to reconstruct. As soon as the reconstruction procedure encounters two subbands with the same 'parent' of which the high-pass subband has shorter length, it knows that a 0 has to be appended on the right-hand side of it, before applying the signal extension in the normal way.

The exact value to which we force 'w' to be equal is not very important for PR. However, the choice will have an effect on the samples in the low-pass band. In general, the statistics of the low-pass band should not change because of our choice for 'w'. Thus, a constant signal should stay constant after adding the specific sample 'h'. It can easily be verified that then the value for 'w' has to be fixed to 0 because of the zero mean of the high-pass filter.

The implementation of the efficient signal extension method does not increase the computational complexity. Instead of the last inner product in the filtering (not needed because it leads to the preset value), we have to calculate the specific sample. This requires the same number of multiplications and additions as for the calculation of an inner product.

2.2.4. Processing of segment lines of length greater than one

The filtering and downsampling process is explained for the rows of the image. The columns are handled in an analogous manner. The row filtering starts at the upper left corner of the image and the rows are scanned from top to bottom, as usual. We do not filter per image row, as in ordinary wavelet filtering, but we filter per segment line, where a segment line is defined to be a horizontal (or, vertical in the case of column filtering) sequence of connected

198

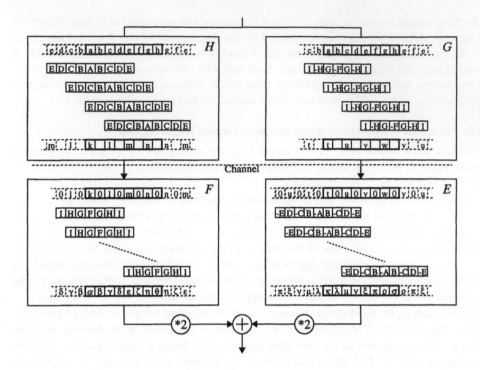

Figure 4   Symmetric extension for symmetric odd length filters and for even length signals.

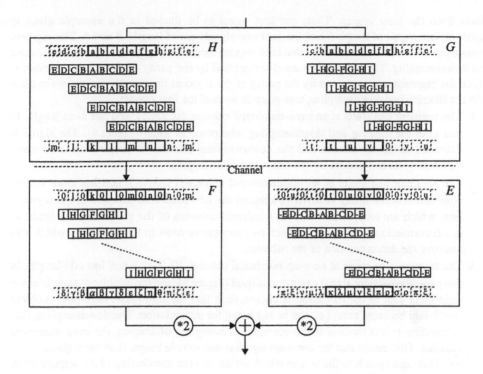

Figure 5   Symmetric extension for symmetric odd length filters and for odd length signals.

pixels from the same region. Thus, our first signal to be filtered in the example given in Figure 3a is a signal of length 6 and the next two signals are of lengths 7 and 5. The segment lines can be divided into four classes that each require their own implementation of the filtering and downsampling. The four classes are distinguished by the parity of the column number at which the segment line starts and by the parity of the segment line length. Below, we explain how the filtering and downsampling take place in each of the four classes.

1. The segment line starts at an even-numbered column $(0, 2, \ldots)$ and has even length. In this case the filtering and downsampling take place as usual (Figure 4). The signal is extended at the boundaries and the downsampling consists of keeping the even-numbered samples and discarding the odd-numbered samples.

2. The segment line starts at an odd-numbered column $(1, 3, \ldots)$ and has even length. Now the downsampling consists of keeping the odd-numbered samples of the segment line, which are located on the even-numbered columns of the row. Then we retain as much correlation as possible between two successive rows in the subbands, which will improve the decomposition of the columns.

3. The segment line starts at an even-numbered column $(0, 2, \ldots)$ and has odd length. In this case, the efficient signal extension is used (Figure 5), and the additional sample at the right-hand side in the extension is chosen such that the right-most high-pass subband coefficient becomes zero, i.e., can be neglected for transmission. The downsampling corresponding to this method is the normal downsampling of keeping the even-numbered samples. This means that the low-pass signal is one sample longer than the high-pass signal. This corresponds to the way in which we defined the transferring of the segmentation to the subbands (Figure 3).

4. The segment line starts at an odd-numbered column $(1, 3, \ldots)$ and has odd length. This is the most difficult case. Now, the resulting low-pass signal must be one sample shorter than the high-pass signal. Further, the downsampling has to take place by keeping the odd-numbered samples of the segment line, which are located on even-numbered columns of the row. This implies that the efficient signal extension has to be implemented in such a way that the additional sample in the extension appears at the *left*-hand side. The determination of this extra sample differs from that of the previous case, because now the pre-set sample is positioned in the low-pass subband. It will be clear that the choice 0 will not do, as it will alter the statistics of the subbands in general. Therefore, we define the pre-set sample in the low-pass subband to be equal to its (right) neighbor. This choice is expected to preserve the statistics of both the low-pass and high-pass signal. Then the calculation of the additional boundary sample requires two inner products of the filter with a part of the signal.

The four classes are illustrated in Figure 6. In each class only the analysis filtering is shown, where H and G are the low-pass and high-pass filter, respectively. The upper line of every block represents the (extended) signal and the lower line the subband. The circles indicate the special samples, whereas the subband coefficients with gray background do not need to be transmitted, i.e., the total number of subband coefficients is equal to the number of pixels in the original signal.

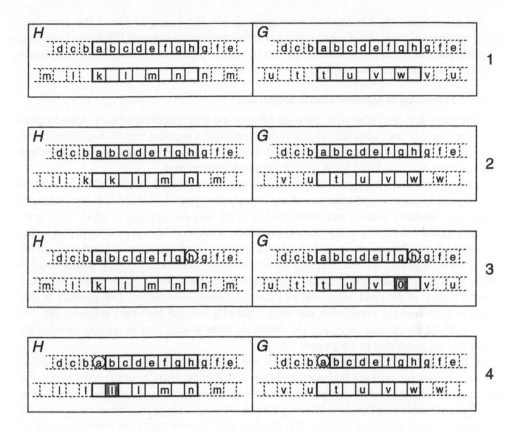

Figure 6   Schematic illustration of the four classes of filtering.

For the columns, the filtering and downsampling procedures are the same. The downsampling depends on the parity of the row number where the vertical segment line starts. Note that, in principle, the signal extension method can be chosen either periodically or symmetrically and that there are no restrictions on the filter class to be used.

2.2.5. Processing of segment lines of length one

We now have described in general terms the filtering and downsampling of a segmented signal. However, there is one exception and this is the situation of a segment line of length one. It will be clear that the information in such a signal of length one is always of low-pass nature and that it cannot be divided into a low-pass and high-pass component.

If this segment line of length one is situated at an even-numbered position, there is no problem. It must be filtered by the low-pass filter to produce a coefficient in the low-pass subband. Thus, it is extended, filtered, and downsampled in the way defined above, which, of course, amounts to just copying the pixel value in the low-pass subband. However, if the segment line of length one is located at an odd-numbered position, it should give a coefficient in the high-pass subband. This coefficient will be zero after the extension, filtering, and downsampling, because of the zero mean of the high-pass filter. As the value of this subband coefficient is independent of the pixel value, we will have lost some information. Our solution to this problem is to keep the pixel value and copy it directly into the high-pass subband. We call these subband coefficients *singles* in the sequel. In other words, singles are low-pass coefficients that are positioned in a high-pass subband.

At this point we meet a difference between row and column processing. We assume that at the encoder first the rows are processed and then the columns. Then, after the row filtering not all segment columns contain low-pass information, because half of the columns are related to the high-pass subband of the row filtering. New singles can only appear when processing the low-pass subband of the row filtering. The coefficients of the row high-pass subbands are all of a high-pass nature and thus a direct copying of a segment part of length one in the column high-pass signal will hardly disturb the statistics of this subband, even when the region is located in an odd pixel position. Therefore, the singles appearing in the HH band are, in fact, singles caused by row filtering only. In Figure 7 we show the singles of the segmented image of Figure 3c.

The slight asymmetry in the row and column processing has the consequence that the reconstruction filtering at the decoder must be performed in reversed order. In our case, this means that the column filtering at the decoder must precede the row filtering, otherwise perfect reconstruction is not possible.

3. THE CODING SCHEME

In this section we describe the image coding scheme used in the experiments reported in the next sections. The scheme is shown in Figure 8. At the encoder, the image is segmented into several regions of similar variance and gray-value. The number of regions depends on the image content and is a compromise between accurate segmentation and transmission costs for

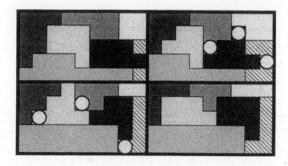

Figure 7 The singles in the decomposition of Figure 3c, indicated by white circles. In this example none of the singles of the row processing falls in the HH band.

the contours. The segmentation controls the discrete wavelet transform of the image, the calculation of the variances, and the quantization. At the decoder, the inverse quantization and inverse transform are controlled by the segmentation as well. Therefore, it is important to have a good segmentation and the segmentation information should be protected against channel errors.

3.1. The segmentation

We use in our experiments the split-and-merge method described in [15]. It starts with a quad-tree splitting of the image up to a certain level. Then each group of four 'child' blocks with the same 'parent' is examined to decide if they should be merged according to the similarity of the gray-value mean and variance inside the blocks. This process is repeated recursively. After the merging step, a recursive splitting process is started. For every block, it is evaluated whether it should be split into four child blocks, depending on the variation in gray-values. Finally, the segmentation is obtained by a grouping procedure that merges blocks from different subtrees if they have similar characteristics.

The segmentation information consists of the separation lines between the regions. These can be transmitted using a 4-connected chain code that follows the links in the separation lines. An alternative method is to use an 8-connected chain code addressing the edge pixels in the regions. In that case the chain is formed by pixels. Note that now not all regions need to be coded with their own edge pixels, because a separation line is completely determined by coding only the chain of pixels on one side of it. In our experiments we did not implement a chain code, but we assumed that the transmission of the segmentation information requires a bit rate of 1.5 bit per link. In [2] we show that this bit rate is an upper bound for a 4-connected chain code.

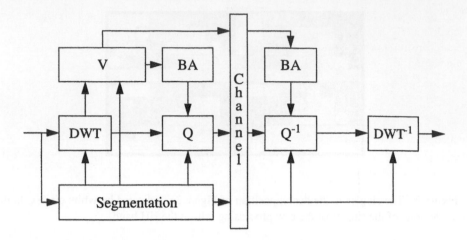

Figure 8   The region-based discrete wavelet transform image coding scheme. The segmentation controls the transform (DWT), the variance calculation (V), and for the low-pass subband also the quantization (Q). The transmitted data consists of the segmentation information, the variances, and the quantizer output. The bit allocation (BA) is repeated at the decoder.

3.2.   The bit allocation

The bit allocation problem consists of finding the optimal distribution of the available bit rate among the subband regions. In our experiments, we used the bit allocation that originates from [16]. If a signal of size N is decomposed into M subband regions, where subband region k has size N_k, then the total bit rate R (in bit per pixel) is

$$R = \sum_{k=0}^{M-1} \frac{N_k}{N} R_k, \tag{1}$$

where the allocated bit rates for the subband regions are denoted by R_k. Each of these allocated bit rates invokes a quantization, which causes a distortion D_k per subband region. We assume that the total distortion of the signal is the sum of the distortions of the subband regions,

$$D = \sum_{k=0}^{M-1} D_k. \tag{2}$$

For each subband region k a set of Q_k quantizers is available, each with its own rate and distortion. These quantizers are located both at the encoder and the decoder side. The bit allocation consists of assigning to each subband region one of its quantizers, in a way such that the total distortion is minimized. Or, if a certain distortion is required, then the bit allocation should find the minimal rate to transmit the signal at the given distortion. The latter case is not examined in this work, as in our application the rate is specified. The number of possible bit allocations is

$$\prod_{k=0}^{M-1} Q_k,$$ (3)

which can be rather large. All these bit allocations can be represented by dots in the rate-distortion plane.

The algorithm consists of finding the optimal distortion, given a target rate, by walking along the lower convex hull of the points until the target rate is achieved. The algorithm is

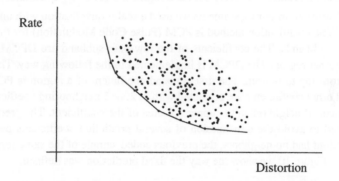

Figure 9   Rate-distortion plane with dots that represent a possible rate-distortion combination. The line is the lower convex hull connecting points with optimal rate-distortion combinations.

computationally attractive, since only a few dots have to be evaluated. For details of the algorithm we refer to [16]. Another advantage of this bit allocation algorithm is that the quantization is left completely free. The only requirement is that the rate and distortion of each quantizer are known. Further, the algorithm can easily be adjusted to give a better adaptation to the human visual system (HVS). If some subband regions have more importance than in our definition, e.g. the regions of the low-pass subband, then only their distortion measures have to be weighted, whereas the allocation procedure remains the same.

The assumption that the total distortion equals the sum of the distortions of the subband regions can be realized by taking the mean square error (MSE) distortion measure and an orthogonal filter pair. The MSE distortion measure does not completely correspond to the working of the HVS, but in the absence of a better distortion measure it is a good choice and it keeps the calculations simple. In the case of non-orthogonal filters, (2) is closely approximated if the energy of the filters is close to 0.5 (within ca. 1%), that is,

$$\sum_{n=0}^{L-1} h_n^2 \approx 0.5, \tag{4}$$

for a filter h_n with length L. If the energy of the filter deviates strongly from 0.5, extra weights have to be introduced, as explained in [18].

The singles are handled separately and the variance calculation and bit allocation of the remaining subband coefficients is performed without the singles. Note that the separate processing of singles is possible because their location is determined by the segmentation mask, which is known both at the encoder and the decoder.

3.3. The quantization

The bit allocation requires for every subband region k a set of Q_k quantizers, each with its own rate and distortion. In our experiments we used a scalar quantization with uniform threshold quantizers. The quantization method is PCM (Pulse Code Modulation) for the coefficients in all high-pass subbands. The coefficients of the low-pass subband are DPCM (Differential PCM) quantized per region. The DPCM is implemented in the following way. The regions are line scanned from top to bottom. The first subband coefficient of a region is PCM quantized with 6 bpp. All other coefficients are predicted with 3, 2, or 1 neighboring coefficients depending on the number of neighbors in the upper left area of the coefficient. The prediction coefficients were fixed to avoid the transmission of several prediction coefficients per region. If a subband coefficient had no neighbors, the previous coded sample of the same region was used as prediction. In Figure 10 we show the way the fixed prediction was defined.

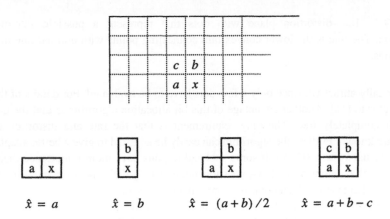

$$\hat{x} = a \qquad \hat{x} = b \qquad \hat{x} = (a+b)/2 \qquad \hat{x} = a+b-c$$

Figure 10   The fixed DPCM prediction $\hat{x}$ of the subband coefficient x in the low-pass subband, depending on the number of neighboring coefficients in the upper left area (denoted by a, b, c) that are located in the same region.

The scalar quantizers were designed to match the statistical properties of the subband coefficients. In the past, quite some research has been dedicated to the determination of the probability density function (pdf) that fits the transform coefficients (e.g. [14]). The generalized Gaussian distribution

$$p(x) = \frac{\alpha K(\alpha)}{\sigma \Gamma(\frac{1}{\alpha})} \, e^{-\left[K(\alpha)\left|\frac{x-\mu}{\sigma}\right|\right]^{\alpha}}, \tag{5}$$

with shape parameter α, scale parameter σ, and location parameter μ has turned out to be the best pdf model for subbands, and we assume the same model for the subband regions. The function $\Gamma(.)$ is the gamma function and the function $K(\alpha)$ is defined as

$$K(\alpha) = \sqrt{\Gamma(\frac{3}{\alpha})/\Gamma(\frac{1}{\alpha})}. \tag{6}$$

If α tends to 0, the shape of the pdf becomes a delta function. For $\alpha = 1$ the Laplacian distribution appears, and for $\alpha = 2$ the Gaussian distribution. If α approaches infinity then the pdf becomes a uniform distribution (see Figure 11). To find the pdf of a certain subband region, the quantization procedure starts with the rescaling of the subband region coefficients

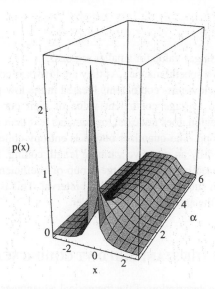

Figure 11 The generalized Gaussian distribution. The pdf is plotted for α between 0.5 and 6.0 whereas the scale parameter σ and the location parameter μ are set to 1 and 0, respectively.

to unit variance. Then the shape parameter is estimated with $\sigma = 1$ and $\mu = 0$ fixed. The choice of zero mean is justified by the zero mean of the high-pass filter. In Figure 12 we show the relation between segment size and shape parameter of the pdf for the Vectra image (Section 4.2). The pdf fitting is based on the minimal Kolmogorov-Smirnov distance, which was evaluated for 13 shape parameters in the range $0.4 - 6.0$. It can be concluded that the

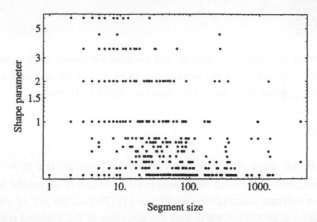

Figure 12   Relation between segment size and shape parameter of the pdf, here shown for the high-pass subbands of a 10-band decomposition of the Vectra image which is segmented into 52 regions (Section 4.2).

shape parameter tends to a larger value for smaller regions.

The singles are coded in a special manner. As they are low-pass coefficients, they can easily be predicted from the other low-pass coefficients located in the low-pass subband. We choose to code the singles by the average coefficient value of their corresponding region in the low-pass subband. If this region does not exist, or contains less than 3 coefficients, we apply a PCM quantization with 6 bpp. The quantizer output is entropy-coded in our implementation. The entropy can be reached by an adaptive variable length coding. For instance, an adaptive arithmetic coding can be used by coding a large group of coefficients simultaneously, where each coefficient has its own subdivision of the [0, 1] interval. This is possible because the pdf models are defined both at the encoder and the decoder.

4.   COMPARISON TO THE STANDARD DWT CODING SCHEME

This section starts with a description of the theoretical advantages and disadvantages of the RBDWT coding scheme in comparison to the standard DWT coding scheme. Next, the coding performances of the new RBDWT coding scheme are evaluated by presenting experimental results. Finally, we examine the ability of the RBDWT to code images hierarchically.

4.1.   Theoretical advantages and disadvantages of the RBDWT coding scheme

In this section we describe the expected advantages and disadvantages of the complete RBDWT coding scheme (Section 3) compared to the standard DWT coding scheme, which was implemented similarly to the RBDWT coding scheme. The only differences are that each subband region corresponds to a complete subband [2] and that the three DPCM prediction coefficients for the low-pass subband are determined by solving the Yule-Walker equations.

There are four advantages of the RBDWT coding scheme in comparison to the standard DWT coding scheme. Firstly, after the segment boundaries have been determined, the filtering takes place inside relatively homogeneous regions. Consequently, the high-pass subbands are expected to contain less energy than for the standard DWT, since filtering across sharp edges is avoided. Another advantage of the RBDWT is that the DPCM coding of the low-pass subband has a better prediction, because the spatial correlation inside each region will be higher than the spatial correlation over the low-pass subband as a whole. Therefore, the energy of the prediction error of the low-pass subband will be lower. In the third place, quantization errors have only local effect, because the reconstruction filtering depends only on one region. That is, quantization errors do not influence neighboring regions and the ringing artifacts near sharp edges will be much less than for the standard DWT. Finally, the bit allocation should be able to favor important regions in the image, as bits can be assigned to each subband region separately.

A disadvantage of the RBDWT coding scheme in comparison to the standard DWT coding scheme is the time-consuming generation of the segmentation. However, the difference in complexity for the filtering process can be neglected, because only one extra inner product is needed for roughly one fourth of the segment lines and columns (class 4 in Figure 6). Another disadvantage is the lower bit rate available for the transform coefficients because of the need for transmitting the segmentation information.

4.2.   Experimental comparison of the RBDWT and the standard DWT coding schemes

In this section we present the coding results of the RBDWT and compare them to the standard DWT. First, the full resolution image coding is considered, and then the multiresolution idea is investigated.

Two test images were used: Camera-man (256x256 pixels) and the first frame of the Vectra video sequence. To facilitate the split-and-merge process with quad-trees we used a 256x256 part of the original Vectra image. Both images were selected because of the presence of sharply defined objects, for which region-based coding methods are most suited.

The segmentation of the Camera-man image used 11 regions, whereas the Vectra image was segmented into 52 regions. These segmentations were selected from several different segmentations, obtained by trial and error using the split-and-merge algorithm of [15]. The choices were made by considering the number of segments and the representation of some objects in the scene. For a complete coding scheme, the segmentation process should be performed automatically, taking into account the texture coding by the RBDWT, that is, relatively large regions. However, here we are mainly interested in the evaluation of the texture coding by the RBDWT, and therefore we did not attempt to optimize the segmentation process. For the same reason, we kept the segmentation mask fixed in all experiments.

210

(a)

(b)

Figure 13   (a) The test image Camera-man; (b) The segmentation of the image into 11 regions; (c) The 10-band decomposition of the segmentation mask; (d) The locations of the 199 singles indicated by black dots.

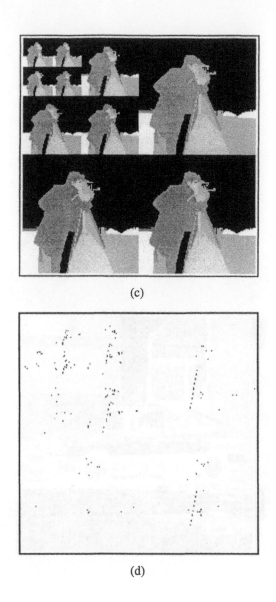

(c)

(d)

Figure 13 (Continued).

(a)

(b)

Figure 14   (a) The 256x256 sized test image Vectra; (b) The segmentation of the image into 52 regions; (c) The 10-band decomposition of the segmentation mask; (d) The locations of the 481 singles indicated by black dots.

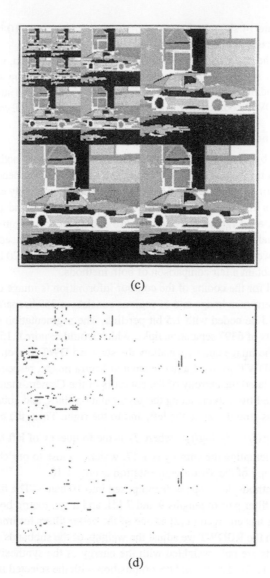

(c)

(d)

Figure 14 (Continued).

Figure 13 and Figure 14 show both images, their segmentation, the 10-band decomposition of the segmentation mask, and the location of the singles.

We continue with the description of the coding scheme settings. Then, we examine the variances of the high-pass subbands and the prediction error variance of the low-pass subband in comparison to the standard DWT. Finally, we compare the objective and subjective coding performances of the RBDWT coding scheme to those of the standard DWT coding scheme.

4.2.1. Coding scheme settings

For bit rates of 1.0 bpp and higher the differences between the two methods will not be visible. For low bit rates that are only slightly above the number of bits needed for transmitting the segmentation information, the costs of the chain code will completely consume the channel capacity, such that for the texture only one quantized low-pass value per region remains. Therefore, we choose a bit rate of 0.20 bpp for the coding of the texture information of the RBDWT. Then, we make an estimation of the additional bits that are needed to code the contour information and we allow the standard DWT to use a bit rate of 0.20 bpp plus those additional bits. Thus, we obtain a fair comparison of both methods.

The bit rate needed for the coding of the contour information is image dependent. Our segmentation of the Camera-man image into 11 regions consists of 2694 separation links, which is 0.06 bpp if they would be coded with 1.5 bit per link. The segmentation of the Vectra image into 52 regions consists of 6397 separation links, which would require 0.15 bpp. Therefore, for a fair comparison of the image quality we allow the standard DWT to spend 0.06 bpp and 0.15 bpp more than the RBDWT, for the Camera-man and Vectra image, respectively.

We have also calculated the entropy of the contours of the Camera-man and Vectra images. This entropy is obtained by walking along the segmentation lines, and counting the number of times the direction was straight on, to the left, and to the right. Then, the entropy is calculated by $-f_S^2 \log(f_S) - f_L^2 \log(f_L) - f_R^2 \log(f_R)$, where f_L is the frequency of left turns, and so on. For the Camera-man segmentation the entropy is 1.47, which is close to our choice of 1.5 bits per link, whereas the entropy of the Vectra segmentation is only 1.10.

For both coding methods the image is decomposed into 10 bands. The filter bank used is the biorthogonal wavelet filter pair of lengths 9 and 7 [4], and a symmetric boundary extension is applied. The filter pair has emerged in [2] as one of the better filters for image coding.

For the coding with the RBDWT we adjust the weights of the subbands in the bit allocation. Normally, all subbands are only weighted with the energy of the synthesis filters [18], which ranges from 0.88 to 1.11 for the 10-band decomposition with the selected filter. However, for a better PSNR and visual quality the low-pass subband of the Camera-man and Vectra image are weighted 20 and 30 times higher than the high-pass subbands, respectively. These weights depend on the average size of the subband regions and have been determined experimentally. For the standard DWT the normal weighting of the low-pass subband emerged as best.

4.2.2. Comparison of the variances

Before considering the objective and subjective coding performance we first compare the subband variances. In Section 4.1 we expected a decrease in variance for the RBDWT with respect to standard DWT.

The variances are calculated assuming a zero mean for each high-pass subband and subband region, which is theoretically correct, because of the zero mean of the high-pass filter. For the low-pass subband and subband regions, we take the variance of the prediction error, again assuming zero mean.

For the RBDWT, we calculate for every subband the average variance of the regions. The average variance for subband s_k with size N_k and variance σ_k^2, which has been divided into r_k segments of size $N_{k,j}$ and variance $\sigma_{k,j}^2$, is calculated by

$$\sum_{j=0}^{r_k-1} (N_{k,j}/N_k) \, \sigma_{k,j}^2 = \sum_{j=0}^{r_k-1} (N_{k,j}/N_k) \left(\frac{1}{N_{k,j}} \sum_{n=0}^{N_{r_j}-1} (s_{k,n})^2 \right) = \sigma_k^2, \tag{7}$$

i.e., just the variance of subband k.

In Table 1 we put the average variances of the subbands of the RBDWT and the subband variances of the standard DWT. We clearly see that for the low-pass subband the prediction error variance has decreased enormously, as had been expected. However, most of the high-pass subbands have an unexpected higher variance (indicated by the italic numbers). The explanation for this effect is most probably that the fourth type of extension reported in Section 2.2 is not preserving the statistics, but is increasing the variance instead. This is caused by the choice of the left-most sample of the low-pass subband to be equal to its right neighbor. Implicitly, this assumes a small variation near the edge and as most edges are not ideal, this will not always be valid. This effect is inherent to our implementation of the RBDWT.

4.2.3. Comparison of the objective and subjective coding performance

In Figure 15a/c and Figure 16a/c we compare the RBDWT with the standard DWT for comparable bit rates, when the contour coding is assumed to require 1.5 bits per link. Two other images are added for comparison. Firstly, in Figure 15b and Figure 16b we added Gaussian noise to the reconstructed image, with a variance that equals a fixed percentage of the variance of the error image (the difference between the original and the reconstructed image). This method has been suggested in [9], to reduce the artifact inherent to region-based image coding schemes which is that the regions seem to be artificially colored ('painted by numbers'). We chose a noise variance percentage of 60, which was determined experimentally. Secondly, the images in Figure 15d and Figure 16d are coded with standard DWT at a bit rate of 0.20 bpp, which would be a fair comparison to the RBDWT if the contour information would not need to be transmitted.

Note that the face in the Camera-man image is reconstructed very well with the RBDWT method (Figure 15a). This accords with our expected advantage of the RBDWT in that it is able to divide its bits not only over the subbands, but also over the spatial domain. Further, the

216

(a)

(b)

Figure 15　Comparison of the coding performance of RBDWT versus DWT. (a) The Camera-man image coded with RBDWT at 0.20 bpp (without counting segmentation overhead of ca. 0.06 bpp); (b) The same image with noise added; (c) and (d) The image coded with the standard DWT at 0.26 and 0.20 bpp, respectively.

(c)

(d)

Figure 15 (Continued).

(a)

(b)

Figure 16 Comparison of the coding performance of RBDWT versus DWT. (a) The Vectra image coded with RBDWT at 0.20 bpp (without counting segmentation overhead of ca. 0.15 bpp); (b) The same image with noise added; (c) and (d) The image coded with the standard DWT at 0.35 and 0.20 bpp, respectively.

(c)

(d)

Figure 16 (Continued).

Table 1 Subband variances for the RBDWT and the DWT

Subband number	Subband variances DWT Camera-man	Subband variances RBDWT Camera-man	Subband variances DWT Vectra	Subband variances RBDWT Vectra
0	509.5	301.2	361.2	82.8
1	105.1	*125.1*	36.9	17.9
2	65.2	*104.6*	209.9	53.9
3	28.7	*101.6*	22.4	21.2
4	112.8	105.8	34.0	17.0
5	48.2	*79.1*	117.2	63.4
6	22.8	*89.9*	11.0	*23.6*
7	69.7	*98.8*	10.6	*21.2*
8	70.8	*93.6*	69.1	*108.0*
9	12.7	*96.0*	4.2	*27.8*

edges are much sharper for the RBDWT and the ringing artifacts are smaller, as had been expected. The standard DWT yields a more natural texture inside the regions. In Figure 15a we see clearly DPCM errors inside the clothes. This seems to suggest that for a small number of regions the use of Yule-Walker prediction coefficients might be preferable to the fixed prediction coefficients defined in Figure 10, which is useful for images with many regions as the Vectra image.

Four test persons evaluated the images. A series of two images were presented and each time they had to choose the one they liked most. If they had no preference at all, they were allowed to make no choice.

Two of the test persons appreciated the noise addition in the images. A cautious conclusion is that if noise is added, the variance should be a small percentage of the variance of the difference image. Further, all test persons ranked the RBDWT images above the images coded with standard DWT at 0.20 bpp.

The most important comparison is the comparison of the RBDWT with the standard DWT at a corresponding bit rate, i.e., 0.26 bpp and 0.35 bpp for the Camera-man and Vectra image, respectively. Both pairs of images were presented twice in the series. The RBDWT coded Camera-man image was twice preferred by all test persons over the standard DWT coded image. However, for the Vectra image two persons chose the standard DWT once. For the remaining choices, the RBDWT coded image was ranked better.

Of course, this experiment does not allow us to reach firm conclusions. However, we may say that the RBDWT is a competitive alternative to the standard DWT in terms of subjective visual quality. The sharpness of the edges and the small ringing artifacts are often considered more important than the lack of texture in some segments.

For the sake of completeness, we show in Figure 17 both images coded with RBDWT, where the low-pass subband is weighted normally. Clearly, the low-pass subbands get too few bits and this is caused by the extremely good prediction of the DPCM inside the segments. The prediction error variances are then so low that the segments in the low-pass subband do not get enough bits.

For comparison of the objective coding performance we put in Table 2 the PSNR of the compressed images of Figure 15, Figure 16, and Figure 17. In all cases the standard DWT has a better PSNR, and the weighting of the low-pass subband for the RBDWT generates a higher PSNR than the normal weighting.

Table 2   Objective coding performance of RBDWT and standard DWT

PSNR (dB)	Texture of RBDWT (0.20 bpp), weighted low-pass subband	Texture of RBDWT (0.20 bpp), non-weighted low-pass subband	Standard DWT (0.26 bpp / 0.35 bpp)	Standard DWT (0.20 bpp)
Camera-man	25.2	24.5	26.4	25.4
Vectra	28.2	27.8	29.3	27.2

4.3.   Multiresolution coding of images using the RBDWT

In some applications, it is a useful feature to build up the image hierarchically using several resolutions. First, a low-resolution image is produced, and when a high-resolution is required, additional information is added to reconstruct the image at medium or full resolution. In this section, we compare the capability of both the RBDWT and the DWT coding scheme to perform such a multiresolution coding, which is useful for the compatible coding of various image formats.

We show in Figure 18 and Figure 19 a straightforward implementation: the original image is decomposed into 10 subbands and only the four subbands of level 3 are transmitted at the first stage. We coded this first stage with 0.07 bpp and did not take into account the coding of the segmentation information. For the RBDWT coding scheme, the reconstructed low-resolution image is shown in Figure 18a and Figure 19a. We clearly see a blocking effect, because of the downsampling of the segmentation. The image coded with standard DWT coding scheme does not suffer from this type of error (Figure 18b and Figure 19b). It shows a normal smoothing of all edges and it has the usual low-pass features. The images coded by the standard DWT scheme are usually appreciated more. It is to be expected that for larger size images the downsampled segmentation will have less effect on the visual quality, which is of importance for HDTV applications.

For the comparison above, we neglected the coding of the segmentation information. However, we need in a hierarchical scheme only the segmentation of the subbands that are transmitted. Therefore, it would be logical to transmit the segmentation per subband, instead of the segmentation of the total image. As the segmentation information consists of (one-dimen-

(a)

(b)

Figure 17　The test images coded with the RBDWT coding method, where the low-pass subband is weighted equally in respect to the high-pass subbands; (a) Cameraman; (b) Vectra.

sional) segmentation lines, the chain codes of all four child subbands in a splitting are roughly *half* the size of the chains codes of the parent subband. This causes a higher total bit rate for the full resolution image. How this increase in total segmentation information can be diminished by using the correlation between the segmentations of the four subbands is an interesting question.

Our overall conclusion is that the multiresolution coding with the RBDWT coding scheme is, at first sight, not competitive to the multiresolution coding with the standard DWT coding scheme. The downsampling of the segmentation causes blocking effects, which grow with the decrease in resolution.

4.4. Application to video coding

The RBDWT can be applied to video coding in several ways. In this section we mention three possibilities and discuss whether these are promising for further research.

The first, trivial, method is to code all frames independently and then the RBDWT can be applied as before. A drawback is that for every frame the segmentation information has to be transmitted, which is quite expensive. As no temporal correlation is used at all, this method will not be optimal. Further, the differences in the segmentation from frame to frame will generate a jerky scene. Therefore, this method is not recommended.

A better method is to estimate the motion of the segments in the scene. If the moving objects are small, probably large parts of the background segmentation will be equal to the segmentation in the previous frame. Thus, the segmentation transmission cost per frame is expected to decrease, because only the modifications to the segmentation have to be transmitted. Research can be directed to find the matching of the segmentations of two successive frames and to code the differences efficiently.

In this approach, the coding of texture information in the regions with RBDWT can be performed in two ways. Firstly, we make use of the temporal correlation before applying the RBDWT. Image regions that have a corresponding (part of a) region in the previous frame can be predicted by the corresponding region (parts) of the previous frame. Then, the prediction error as well as the non-predicted areas can be coded by the RBDWT. Secondly, we can think of using the temporal correlation after applying the RBDWT. However, the motion estimation between the coefficients of subband regions is even more complex than for subbands as a whole. The reason is that the shapes of the segments depend on the position of the corresponding region in the frame (compare the different decompositions of the two central segments in Figure 3). Because the usual aliasing problems of motion estimation in subbands also show up, this type of coding scheme is not expected to be useful. The first method of coding the texture is expected to be the most promising. Further research must indicate whether it is competitive with existing video coding schemes or not.

A third method has been suggested in [13] and used in [8] for the application of the generalized orthogonal transform to video coding, but it can also be used for the RBDWT. A new frame is predicted from the previous one, thus without using any segmentation information. From the prediction error image the areas are determined for which the motion compensation has failed. The pixels that are masked by these areas have a high-pass characteristic, i.e., the

(a)

(b)

Figure 18   Reconstructed Camera-man images using only the subbands of level 3 of a 10-band decomposition, coded at 0.07 bpp. The images have been enlarged by factor 4. (a) RBDWT; (b) DWT.

(a)

(b)

Figure 19   Reconstructed Vectra images using only the subbands of level 3 of a 10-band decomposition, coded at 0.07 bpp. The images have been enlarged by factor 4. (a) RBDWT; (b) DWT.

correlation between the pixels is rather low and intra-frame coding methods like the RBDWT are not useful. Instead, the corresponding regions in the original frame are coded by the RBDWT.

From the three possible methods mentioned above, the third method seems to be the most promising, since it does not make any assumptions about the correspondence between the segmentations of two consecutive frames.

5. CONCLUSIONS AND FURTHER RESEARCH

In this chapter, we have presented a new texture coding method for region-based coding techniques which is called the region-based discrete wavelet transform (RBDWT). It has the advantage that it is able to distribute the available bit rate over subbands as well as regions. We have presented a coding scheme, which was used to code two test images showing sharply defined objects.

The subjective and objective coding performance of the RBDWT have been compared with the standard DWT. Although the standard DWT has a higher PSNR than the RBDWT, the subjective coding performance of the RBDWT has emerged to be quite competitive. In particular, the edges are sharper and ringing artifacts are smaller. For multiresolution coding, the low-resolution image has shown a lower subjectively rated visual quality for the RBDWT than for the standard DWT. For the application of the RBDWT to video coding, the most promising option seems to be the coding of those areas for which motion compensation has failed.

Further research should be directed to the segmentation process, taking into account the characteristics of the RBDWT method for texture coding. Then a complete coding scheme can be built, the coding performance of which should be compared to other region-based coding techniques. Another part of the coding scheme that deserves more attention is the quantization of the subband coefficients. Some gain is to be expected by using another pdf model for small-sized regions. Another research area is the development of an efficient hierarchical transmission of the segmentation information, which makes the coding of the subbands less dependent on the segmentation of the original image. Finally, the possible application of the RBDWT in a video coding scheme deserves further research.

BIBLIOGRAPHY

1   M. Antonini, M. Barlaud, P. Mathieu, and I. Daubechies, "Image Coding Using Wavelet Transform," *IEEE Transactions on Image Processing*, Vol. 1, No. 2, pp. 205-220, April 1992.

2   H. J. Barnard, *Image and Video Coding Using a Wavelet Decomposition*, Ph. D. thesis, Delft University of Technology, Department of Electrical Engineering, Information Theory Group, Delft, The Netherlands, May 1994.

3   H. J. Barnard, J. H. Weber, and J. Biemond, "Efficient Signal Extension for Subband/-Wavelet Decomposition of Arbitrary Length Signals," *SPIE Vol. 2094 Visual Communications and Image Processing '93*, pp. 966-975, November 1993.

4   A. Cohen, I. Daubechies, and J. -C. Feauveau, "Biorthogonal Bases of Compactly Supported Wavelets," *Communications on Pure and Applied Mathematics*, Vol. XLV, pp. 485-560, 1992.

5   M. Eden, and M. Kocher, "On the Performance of A Contour Coding Algorithm in the Context of Image Coding; Part 1: Contour Segment Coding," *Signal Processing*, Vol. 8, pp. 381-386, 1985.

6   M. Eden, M. Unser, and R. Leonardi, "Polynomial Representation of Pictures," *Signal Processing*, Vol. 10, No. 4, pp. 385-393, 1986.

7   H. Freeman, "On the Encoding of Arbitrary Geometric Configurations," *IRE Transactions on Electronic Computers*, Vol. EC-10, No. 2, pp. 260-268, 1961.

8   M. Gilge, T. Engelhardt, and R. Mehlan, "Coding of Arbitrarily Shaped Image Segments Based on a Generalized Orthogonal Transform," *Signal Processing: Image Communication*, Vol. 1, No. 2, pp. 153-180, October 1989.

9   M. Kunt, A. Ikonomopoulos, and M. Kocher, "Second-Generation Image Coding Techniques," *Proceedings of the IEEE*, Vol. 73, No. 4, pp. 549-574, April 1985.

10   M. Kunt, M. Bénard, and R. Leonardi, "Recent Results in High-Compression Image Coding," *IEEE Transactions on Circuits and Systems*, Vol. CAS-34, No. 11, pp. 1306-1336, November 1987.

11   S. G. Mallat, "A Theory for Multiresolution Signal Decomposition: The Wavelet Representation," *IEEE Transactions on Pattern Analysis and Machine Intelligence*, Vol. 11, No. 7, pp. 674-693, July 1989.

12   S. A. Martucci, "Signal Extension and Noncausal Filtering for Subband Coding of Images," *SPIE Vol. 1605 Visual Communications and Image Processing '91: Visual Communication*, pp. 137-148, 1991.

13   H. G. Musmann, M. Hötter, and J. Ostermann, "Object-Oriented Analysis-Synthesis Coding of Moving Images," *Signal Processing: Image Communication*, Vol. 1, No. 2, pp. 117-138, 1989.

14   R. C. Reininger, and J. D. Gibson, "Distributions of the Two-Dimensional DCT-Coefficients for Images," *IEEE Transactions on Communications*, Vol. COM-31, No. 6, pp. 835-839, June 1983.

15   K. C. Strasters, and J. J. Gerbrands, "Three-Dimensional Image Segmentation Using a Split, Merge and Group Approach," *Pattern Recognition Letters*, Vol. 12, No. 5, pp. 307-325, May 1991.

16   P. H. Westerink, *Subband Coding of Images*, Ph. D. thesis, Delft University of Technology, Department of Electrical Engineering, Information Theory Group, Delft, The Netherlands, October 1989.

17   J. W. Woods, and S. D. O'Neil, "Subband Coding of Images," *IEEE Transactions on Acoustics, Speech, and Signal Processing*, Vol. ASSP-34, No. 5, pp. 1278-1288, October 1986.

18   J. W. Woods, and T. Naveen, "A Filter Based Bit Allocation Scheme for Subband Compression of HDTV," *IEEE Transactions on Image Processing*, Vol. 1, No. 3, pp. 436-440, July 1992.

Wavelet transform and motion estimation for image sequence coding: a multiconstraint approach

Nadia BAAZIZ, Claude LABIT

IRISA/INRIA, Campus de Beaulieu 35042 Rennes Cedex, France
e-mail: labit@irisa.fr

Abstract

This paper investigates the efficiency of several multigrid motion estimation schemes to improve image sequence coding using motion compensated predictions. Many previous studies proved the interest to design hierarchical and compact representation for image data (lowpass pyramids, QMF and/or wavelet decompositions...) applied to one spatial image and, moreover, some numerous experiments and codec designs have already used motion compensated prediction loops. The objectives of this paper are to take advantage of a hierarchical decomposition to build efficient motion estimator simultaneously in several frequency bands. Usual differential estimation methods are experimented on wavelet pyramids with several merging operators. A modified Wiener-based motion estimator is presented . It improves the motion estimation efficiency within an orthogonal oriented image decomposition. Finally, results are shown and compared on television image sequences.

1 INTRODUCTION

Multiresolution representations have been fruitfully introduced in image processing and image data compression; indeed they perform much better in comparison with the usual transform techniques (FFT,DCT ...) which are poor in spatial-frequency localization and introduce blocking effects. Multiresolution transforms are also known as subband decomposition techniques or pyramid transforms [1], [2], [3]. Each of them have been introduced separately and formulated with different mathematical tools. The wavelet transform and the associated multiresolution analysis as it has been introduced [4], [5], [6] relate all these techniques by associating a mathematical framework giving efficient modelling and interpretation tools for space-frequency pyramidal decomposition.

Our study ([7], [8]) mainly investigates how all these pyramidal representation schemes work with spatio-temporal (i.e 2-D+t) data information (i.e image sequences indexed by two (x,y) spatial components and one temporal axis) ; when such a spatial hierarchical representation is applied, we can observe along the temporal axis that the visual appearance of motion and spatio-temporal structures are quite interpretable and significant in terms of correlated motions within different somewhat decorrelated frequency subbands. The general goal of this study is to derive numerically this visual coherence of motions when a pyramidal representation is used as the basic data structure. Obviously the return of motion information is of great interest for coding purpose, when a motion compensated prediction loop is introduced in an interframe predictive coding scheme to obtain the best redundancy removal and associated bit rate reduction when a transmission service is involved [9][10][11].

This paper only describes the multigrid motion estimation scheme that we can apply to a motion compensated coder and its efficiency in terms of motion estimation errors. We don't focus herein on the applications to explicit motion compensation coding schemes (see [8] and [12] for these issues) .

Section 2 briefly presents a unified classification of several multiresolution representation methods which enable the design of pyramidal structures for image data. These methods have been introduced, sometimes, independently and sequentially. We compare all of them using the same criteria of entropy along a real TV image sequences. The main goal of this section is to prove the efficiency of orthogonal pyramidal decompositions to compress data representations. Then, we summarize in section 3 the general framework that we use, for motion estimation : it is based on differential methods which can be easily extended to hierarchical representations. Such a new algorithmic extension is proposed and tested.This algorithmic framework is firstly tested on monoresolution image sequences. The sections 4 and 5 present new motion estimation techniques which exploit the basic hierarchical data structure due to any pyramidal decomposition previously introduced. These techniques are defined as multigrid motion estimation (section 4) and denoted by multiconstraint algorithms (section 5) if several frequency subbands are simultaneously processed. Finally some results are shown on real image sequences and motion estimation errors are evaluated

2 MULTIRESOLUTION REPRESENTATIONS OF IMAGE SEQUENCES

2.1 A generalized typology of multiresolution decompositions

The main objective of this section is to present a synthesized and classified typology of several (already known) pyramidal decomposition techniques which have been, in the past introduced separately and not really compared on a unique experimental set of image data and criteria. Especially, for image coding applications, we look at these techniques as pyramid transforms based essentially on filtering and decimation processes.

The pyramid levels are equivalent to frequency channels with adapted bandwith and orientations. Figure 1 gives our typology tree and associated references. It is based essentially on the following classification criteria :

- Filtering properties : we can separate first of all pyramid transforms based on band-pass filtering (Laplacian pyramid, Pseudo-Laplacian pyramid, subband decompositions...) from those based only on lowpass filtering which provide us with several approximated versions of the original image data at successive resolutions (Gaussian pyramid, Pseudo-Gaussian pyramid, Crible).

- Orthogonality : within the family of bandpass pyramids, we distinguish three sub-classes which introduce respectively a non-orthogonal representation (Laplacian pyramid, Pseudo-Laplacian pyramid, Crible), a quasi-orthogonal one which means that only an orthogonality approximation is achieved, and finally strictly orthogonal representation.

- Separability : one usual method consists to extend the use of filters designed for monodimensional (1-D) signals to multidimensional signals (2-D or more than) in a separable way. However, there are some studies which have introduced non-separable bandpass filtering which could isolate some more isotropic patterns within successive pyramidal levels.

All the references corresponding to these classification criteria are given in figure 1. The following subsections present some technical aspects of implementation for a subset of such decompositions. We focus our attention on some representative pyramid transforms that we use in order to dress a comparison table of their performances. Different criteria are etablished to evaluate the decomposition efficiency such as the implementation complexity and spatio-frequency localization. In the point of view of data compression, we compute partial and global entropies to estimate the degree of information decorrelation obtained for each decomposition.

Since we have 3-D ((x, y, t)) data information corresponding to television image sequence, the purely spatial transforms are applied to each field (or frame if progressive signals are involved) to obtain a pyramid sequence.

Two sequences, especially taken for their representative contents of structures and motion features, have been used for the simulations presented here. The first sequence called "INTERVIEW"(by courtesy of BBC-UK) is a real television sequence of 32 frames. Two successive frames of it and the corresponding temporal difference are shown in figure 11. We show respectively in figure 12 the second sequence called "RUBIX" (by courtesy of CCETT-FRANCE) which is an artificial sequence of 32 frames too.

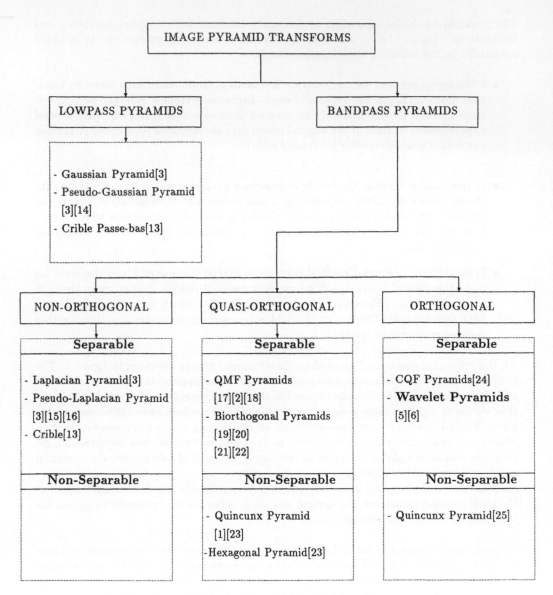

Figure 1: Typology tree of image pyramid transforms

2.2   Comparison based on local and global subband entropies

2.2.1   Laplacian or pseudo-laplacian pyramids

Since lowpass pyramids are very redundant data structures, one can derive bandpass pyramids that captures only the information difference between two successive lowpass levels. If a simple difference operation is performed, the reduced level must be interpolated to have the same size than the highest resolution one. One well-known non-orthogonal bandpass pyramid was introduced by Burt and Adelson in 1983 [3] using a Gaussian pyramid. It was called the Laplacian pyramid since the difference of Gaussian kernels gives an approximation of the Laplacian of the Gaussian. For other lowpass kernels, the obtained pyramid will be called a pseudo-Laplacian pyramid.

N Laplacian or pseudo-Laplacian pyramid levels $(L_0, L_1, ..., L_{N-1})$ are given by :

$$G_{j+1}(k,l) = \sum_n \sum_m h(n)h(m)G_j(2k - m, 2l - n) \tag{1}$$

$$\tilde{G}_{j+1}(k,l) = 4 \sum_n \sum_m h(n)h(m)G_{j+1}(\frac{k+m}{2}, \frac{l+n}{2}) \tag{2}$$

$$L_j(k,l) = G_j(k,l) - \tilde{G}_{j+1}(k,l) \tag{3}$$

$$L_{N-1}(k,l) = G_{N-1}(k,l) \tag{4}$$

$$-M \leq m, n \leq M$$
$$0 \leq j \leq N - 2$$

where G_0 is the original image and $\tilde{G}_j$ the interpolated version of a lowpass pyramid level G_j. For the filtering process, the signal may be extended out of its finite support in the simplest way in order to have easy computation.

The generating kernels can not be ideal lowpass filters, then the difference operation conserves some low-frequency energy and makes the different pyramid levels partially redundant. In other words, the transform is not orthogonal and consequently, as in the lowpass pyramid, the amount of the obtained data structure attains the 4/3 of the original image size.

Partial and global entropy computation performed on pyramid levels shows the information decorrelation obtained by this pyramidal transform and outlines its advantages for coding purposes.We consider that entropy values measure asymptotically the remaining amount of innovative information after any decorrelation attempt.
The global entropy E_g is given by the average of the pyramid level entropies E_{L_j} and represents the optimal number of bits required per pyramid sample in comparison with an original sample :

$$E_g = \sum_{j=0}^{N-1} \frac{E_{L_j}}{2^{2j}} \tag{5}$$

The entropy E of an image is defined as :

$$E = -\sum_i p_i log_2(p_i) \tag{6}$$

where p_i is the distribution probability of its pixel values i.

The following table gives original, partial and global entropy values for :

- "RUBIX" frame and its 3 level Laplacian pyramid generated with 5-tap Gaussian filter.

- " INTERVIEW" frame and its 3 level pseudo-Laplacian pyramid generated from 9-tap lowpass filter (see figure 13).

RUBIX frame		INTERVIEW frame	
Original	3.84	*Original*	7.51
E_{L_0}	3.31	E_{L_0}	3.83
E_{L_1}	3.43	E_{L_1}	4.44
E_{L_2}	4.01	E_{L_2}	4.47
E_g	4.42	E_g	5.22

It could be noticed that partial entropies have small values; this effect is quite expected and suitable for coding applications. However, these transforms suffer from their non-orthogonality properties which make an increase of the data amount and higher global entropy than in the orthogonal case.

2.2.2 Orthogonal bandpass pyramid transforms

Orthogonal pyramid transforms correspond essentially to exact reconstruction subband coding schemes which split the image spectrum into N uncorrelated frequency channels by bandpass filtering and adapted subsampling. The original signal can be exactly reconstructed by inverse operations.

The analysis/synthesis filters have the fondamental property to form an orthonormal set. Conjuguate mirror filters (CQF) [24] and wavelet filters [5] are examples of filter banks which generate this kind of pyramids.

Using wavelet filter banks, the original image is decomposed in a set of subimages corresponding to a dyadic resolution sequence (2^{-j}) [6]. This transform is implemented using a pair of non-symmetrical analysis/synthesis filters (H, G) derived from orthonormal wavelet basis with compact support [5]. They are 1-D finite impulse response filters (FIR) which leads to orthogonality and exact reconstruction concepts as it can be seen through the following equations:

$$g(n) = (-1)^n h(-n+1) \text{ : conjuguate mirror relation.} \tag{7}$$

$$\sum_n h(n) = 1 \text{ : h is a normalized lowpass kernel.} \tag{8}$$

$$\sum_n g(n) = 0 \text{ : g is a highpass kernel.} \tag{9}$$

$$\sum_n h(n-2k)g(n-2l) = 0 \quad : \text{orthogonality relation.} \tag{10}$$

$$\sum_k h(m-2k)h(n-2k)+g(m-2k)g(n-2k) = \delta_{mn} \quad : \text{exact reconstruction relation.} \tag{11}$$

$$\bar{h} = h(-n) \; ; \bar{g} = g(-n). \tag{12}$$

One decomposition step gets four subimages corresponding to three oriented high frequency bands and a low frequency band. The following equations show how to combine the filters and the decimation process to get the four subbands or the first pyramid levels.

$$HH_1(k,l) = \sum_n \sum_m \bar{g}(2k-n)\bar{g}(2l-m)I_0(n,m) \tag{13}$$

$$HB_1(k,l) = \sum_n \sum_m \bar{g}(2k-n)\bar{h}(2l-m)I_0(n,m) \tag{14}$$

$$BH_1(k,l) = \sum_n \sum_m \bar{h}(2k-n)\bar{g}(2l-m)I_0(n,m) \tag{15}$$

$$BB_1(k,l) = \sum_n \sum_m \bar{h}(2k-n)\bar{h}(2l-m)I_0(n,m) \tag{16}$$

I_0 : original image.
$\bar{h}$: impulse response of lowpass wavelet filter.
$\bar{g}$: impulse response of highpass wavelet filter.
HH_1 : subimage of diagonal high frequencies, isolates diagonal edges.
HB_1 : subimage of horizontal high frequencies, isolates vertical edges.
BH_1 : subimage of vertical high frequencies, isolates horizontal edges.
BB_1 : subimage of low frequencies, represents the context of the image.

For boundary filtering, the image signal is assumed to be periodic.

The decomposition is applied recursively to the low frequency band to obtain the other levels. The whole decomposition can be seen as three oriented N level pyramids $(HH_j, BH_j, HB_j)_{j=1...N}$ with a residual frequency band BB_N. The number of transform samples is equal to the original number because of the orthogonal property .

Figure 2 shows the grid representation of the wavelet pyramids in the space domain and the corresponding frequency partition.

The original signal can be perfectly recovered by adding the different interpolated pyramid levels to the residual frequency band (pyramid top). The reconstruction algorithm for 1 level pyramids is described by the following equations :

$$I_0 = \tilde{B}B_1 + \tilde{H}H_1 + \tilde{H}B_1 + \tilde{B}H_1 \tag{17}$$

$$\tilde{B}B_1(n,m) = 4\sum_k \sum_l h(n-2k)h(m-2l)BB_1(k,l) \tag{18}$$

$$\tilde{H}H_1(n,m) = 4\sum_k \sum_l g(n-2k)g(m-2l)HH_1(k,l) \tag{19}$$

$$\tilde{H}B_1(n,m) = 4\sum_k \sum_l g(n-2k)h(m-2l)HB_1(k,l) \tag{20}$$

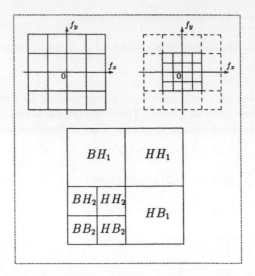

Figure 2: Spatial representation grid of a 2 level wavelet pyramids and the corresponding frequency partition.

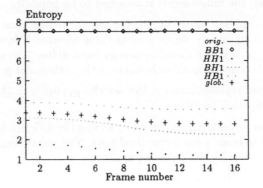

Figure 3: Original, partial and global entropy values computed on 16 "INTERVIEW" frame sequence and on the corresponding 1 level wavelet pyramid sequence.

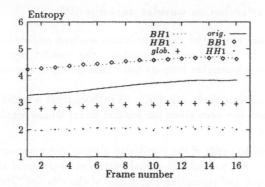

Figure 4: Original, partial and global entropy values computed on the 16 odd "RUBIX" field sequence and on the corresponding 1 level wavelet pyramid sequence.

$$\tilde{B}H_1(n,m) = 4 \sum_k \sum_l h(n-2k)g(m-2l)BH_1(k,l) \tag{21}$$

where $\tilde{B}B_1, \tilde{H}H_1, \tilde{H}B_1 and \tilde{B}H_1$ denote the interpolated versions of $BB_1, HH_1, HB_1 and BH_1$.

In order to get an idea of the degree of the information decorrelation obtained by this decomposition, we perform a global entropy computation as follows: for a given N level wavelet pyramids, we compute the average of subimage entropies E_i to get the optimal number of bits required per pyramid sample by comparison with an original sample.

$$E_g = \sum_{j=1}^{N} \frac{E_{HH_j} + E_{HB_j} + E_{BH_j}}{2^{2j}} + \frac{E_{BB_N}}{2^{2N}} \tag{22}$$

Original, partial and global entropy values are given in the table below for "INTERVIEW" and "RUBIX" frames and their 2 level wavelet pyramids generated from 4-tap wavelet filters [5] (see figure 14 and 15). Figure 3 and 4 show the stability of these values along the pyramid sequences. We observe a significant decrease of the global entropy measures between the non-orthogonal (Laplacian pyramids) and the orthogonal (wavelets) decompositions. These entropy results are in favour of the second ones (more than 2bits/pels) and tends to choose these representations inside motion-based coding schemes.

RUBIX frame		INTERVIEW frame	
Original	3.84	Original	7.51
E_{BB_2}	4.10	E_{BB_2}	7.47
E_{BH_2}	3.12	E_{BH_2}	3.74
E_{HB_2}	2.34	E_{HB_2}	3.84
E_{HH_2}	2.26	E_{HH_2}	2.87
E_{BH_1}	2.88	E_{BH_1}	2.70
E_{HB_1}	1.89	E_{HB_1}	3.65
E_{HH_1}	1.80	E_{HH_1}	1.42
E_g	2.38	E_g	3.06

2.2.3 Regularity criterion on wavelet decomposition

The wavelet theory and associated multiresolution analysis enable to introduce the concept of regularity of the approximation operator ; the wavelet regularity is relative the infinitely iterated and subsampled lowpass filter that needs to converge to some smooth function .

I. Daubechies has investigated the regularity of orthonormal wavelet with compact support and a construction has been given for wavelet filters whose regularity increases with the support length [5].

We are here interested to verify if the regularity give us some improvement in the information decorrelation . So, partial and global entropies of pyramid sequences have been computed with different and increasing regularity of the wavelet filters ; then it appears that entropy values are closely similar (see figure 5). In conclusion we could say that there is no significant improvement of entropy criterion when filter regularity increases in comparison with the increase of the computational complexity due to filter support length. This partial conclusion could be obviously contradicted if any other comparison criterion - such as robustness of the decomposition to quantization errors - is discussed.

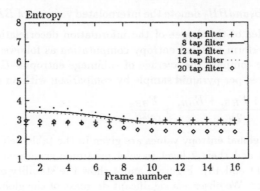

Figure 5: Global entropy values computed at 2 level wavelet pyramids of "INTERVIEW" sequence and obtained for different tap wavelet filters with increasing regularity.

2.3 Discussion

Let us remember that all the pyramid representations discussed here have been previously introduced in the litterature; however, we thought useful to compare them using a same set of criteria, and summarize them within a unique typology. The main interests of this experimental part is to observe that some pyramidal decompositions provide us with highly compact representation (especially for wavelet pyramids), that those performances are quite stable during a temporal sequence and among several contents of scene. It is one of the main reasons why motion estimation and compensation schemes using wavelet decompositions have to be explored.

Moreover, if we pay attention to the coherent visual appearance of moving structures when all of these pyramidal image decompositions are seen in real time along a large TV sequence, visual motion informations appear coherent from one frequency subband to the others; so,it seems quite attractive to numerically quantify this visual coherence in terms of motion estimations. The following section gives some recalls on the general framework for motion estimation that we propose and subsequent sections describe new motion estimation schemes well fitted to pyramidal data structure and motion compensated coding purpose. We will essentially use in the following sections orthogonal band-pass pyramids based on wavelet decomposition ; however, all of these motion estimation studies could be applied in a more general sense to all pyramidal decompositions introduced previously.

3 MONORESOLUTION MOTION ESTIMATION FRAMEWORK

The general framework for motion estimation that we experimented here is based on well-known pel-recursive differential techniques [9], [10], [11]. The basic assumption relating these methods is the temporal invariance of the displaced pel luminances and, in a general way, the recursive estimation procedure can be seen as a minimization of an interframe reconstruction quality criterion. In this paper, we propose an improved recursive procedure which combines the Walker-Rao and the Wiener-based approaches for motion estimation. However, a review of these techniques is of necessity to introduce the modified one.

3.1 Adaptive pel-recursive motion estimation algorithm

This algorithm was introduced by Walker and Rao in 1984 [10]. It uses a steepest descent method to find, for each moving pel p, the optimal motion vector $\hat{d}$ which minimizes the squared displaced frame difference $(DFD)^2$. Let $I(p, t)$ be the intensity value at location $p = (x, y)$ of a moving pel from frame $t - 1$ to frame t. Then its DFD is defined as :

$$DFD(p, \hat{d}) = I(p, t) - I(p - \hat{d}, t - 1) \tag{23}$$

where $\hat{d}$ is its estimated interframe displacement and is given by the following iterative equation :

$$\hat{d}^i = \hat{d}^{i-1} - \Gamma DFD(p, \hat{d}^{i-1}) grad I(p) \tag{24}$$

The update can be controlled by an adaptive gain :

$$\Gamma = \frac{1}{2} \frac{1}{||grad I(p)||^2} \tag{25}$$

Note that i represents the iteration counter and $grad I(p) = (g_x(p), g_y(p))$ the spatial gradient of the luminance I at location $(p - \hat{d}^{i-1})$ in frame $t - 1$. The convergence is achieved when $|DFD(p, \hat{d}^i)| \leq T$ where T is a given threshold value.

3.2 Wiener-based algorithm

The Wiener-based approach for motion estimation was proposed by Biemond and al. [11]. It is a pel recursive technique which provides with a linear least square estimate of the motion vector, for moving pel, using some observations extracted from a spatial neighbourhood Ω. Assuming that a set Ω of N pels $\Omega = \{(p(1), ..., p(j))_{j=1,2,...,N}\}$ belongs to a moving area, then one can derive N observations by linearizing the DFD function for each pel at the location $(p(j) - d^{i-1})$:

$$DFD(p(1), d^{i-1}) = gradI(p(1))u + v(p(1), d^{i-1}) \tag{26}$$

$$\vdots$$

$$DFD(p(N), d^{i-1}) = gradI(p(N))u + v(p(N), d^{i-1}) \tag{27}$$

where $v(p(j), d^{i-1})$ denotes the truncation error resulting from the linearization and $u = (d - d^{i-1})$ the update of the previous estimate vector d^{i-1}. Both u and v are considered as samples of stochastic processes. Estimating u is stated as a problem of finding a linear estimator L such that $\hat{u} = Lz$ and $E\{||u - \hat{u}||^2\}$ minimized. In addition of this, u and v are assumed to be orthogonal each other with zero mean value and diagonal covariance matrixes. In this way, Biemond et al. obtain an iterative algorithm to estimate a current pel displacement using a neighbourhood information as described by the following equations :

$$\hat{d}^i = \hat{d}^{i-1} - \left[\begin{array}{cc} \sum_j g_x^2(j) + \mu & \sum_j g_x(j)g_y(j) \\ \sum_j g_x(j)g_y(j) & \sum_j g_y^2(j) + \mu \end{array} \right]^{-1} \tag{28}$$

$$\left[\begin{array}{c} \sum_j g_x(j)DFD(p(j), \hat{d}^{i-1}) \\ \sum_j g_y(j)DFD(p(j), \hat{d}^{i-1}) \end{array} \right]$$

$$1 \leq j \leq N$$

where μ is equal to the ratio of the variance of v and u ($\mu = \frac{\sigma_v^2}{\sigma_u^2}$) and $(g_x(j), g_y(j)) = gradI(p(j))$ represents the spatial gradient of the luminance I at location $(p(j) - \hat{d}^{i-1})$ in frame $t - 1$.

3.3 Modified Wiener-based approach

If we consider the equations (24) and (28), one can notice that the Wiener algorithm becomes a usual Walker-Rao estimator if the neighbourhood Ω is restricted to the current pel $N = 1$ and $\mu = 0$. Moreover, the computational structure is quite similar between a Walker-Rao iteration and a Wiener-based one; so, some architecture modules could be duplicated if they are used together. Both equations (24) and (28) are frequently used with identical logic operators. However, the Wiener-based approach works better when the current pel and its neighbourhood belong to a moving object but fails when the neighbourhood system goes through motion discontinuities and very noisy areas ; therefore, it requires some estimation process without any smoothing effect such as Walker-Rao one.

To solve this problem, an improved estimation procedure is developped ; it combines a Wiener-based and a Walker-Rao iterations. One combined iteration step (i) is shown by the diagram in figure 6 and can be described as follows :

- Based on an previously estimated displacement $\hat{d}^{i-1}$ of the current pel p, simultaneously a Wiener-based update and a Walker-Rao one are evaluated to get respectively a new displacement v^i and w^i.

- The optimal motion estimate is a-posteriori selected on a motion estimation error criterion (some others criteria could be herein applied to force the efficiency of one of the two extimators). This test procedure can be seen as a tentative to define an adaptive neighbourhood system oscillating between a single point and a larger clique .

- Iterations stop when the current pel is compensated i.e $|DFD(p, \hat{d}^i)| \leq T$; however, because it still exists some cases for which there is no convergence, the iteration process is restricted to an a-priori maximal number of iterations denoted by $imax$.

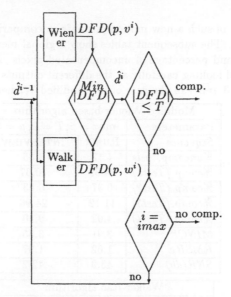

Figure 6: Diagram of the combined Wiener-Walker iteration procedure.

Moreover, it should be mentioned that this iterative motion estimation procedure is one part of the whole motion estimation process which proceeds on a pel-by-pel along the scan direction, as described by the following algorithm:

1. An initial displacement value d^0 is needed to start the iterative process at pel p. It will be one of the displacements found at previous pels, in a causal neighbourhood Ω_{iv}, which gives the smaller $DFD(p, d^0)$ at the current pel. Note that the window

of initial values Ω_{iv} is not necessary the same as Ω.

2. *Test1-* If $|DFD(p, d^0)| \leq T$ then $\hat{d} = d^0$. The correct motion displacement is found only by the simple propagation of the previous displacement.

3. *Test2-* Else if $I(p, t) = I(p, t-1)$ then $\hat{d} = 0$. The current pel is not moving.

4. *Test3-* Else use the combined Wiener-Walker procedure to estimate the correct displacement. If $\sum_{j=1}^{N} |DFD(p(j), d^0)| \leq \sum_{j=1}^{N} |DFD(p(j), 0)|$ then use d^0 to start the iterative procedure else start the iterative procedure with zero value. The iteration process goes on until $|DFD(p, \hat{d}^i)| \leq T$ (convergence) or $i = imax$.

5. *Test4-* If the modified Wiener procedure terminates without any convergence then use a simple Walker-Rao procedure with the initial value d^0 to estimate the correct update.

We test the efficiency of such a new motion estimate in comparison with the reference of Walker-Rao estimate. The subsequent tables show a global measure of its performances in terms of RMSRE and percentage of uncompensated pixels. A more detailed analysis can be also performed looking carefully at the different outputs of the intermediate tests (sequentially test 1,2,3 and 4 in the case of the modified Wiener algorithm).

Modified Wiener-based algorithm		
Parameters	$imax = 5, T = 2, \mu = 5$	
Sequence	RUBIX	INTERVIEW
%uncomp. pels	4.12	3.85
%comp.(Test1)	84.39	70.37
%comp.(Test2)	0.37	0.83
%comp.(Test3)	11.12	24.96
%comp.(Test4)	4.02	9.10
MINCP	3.91	3.55
RMSRE	1.62	1.97
SNR(db)	43.9	42.2

Walker-Rao algorithm		
Parameters	$imax = 5, T = 2$	
Sequence	RUBIX	INTERVIEW
%uncomp. pels	10.29	18.48
%comp.(Test1)	81.96	64.18
%comp.(Test2)	0.27	0.74
%comp.(Test3)	7.48	16.60
MINCP	2.13	1.93
RMSRE	4.96	3.45
SNR(db)	34.2	37.4

Several parameters have to be adjusted ; for example, we have to design the neighbourhood systems Ω and Ω_{iv}, μ and the threshold value T. Simulations were done using several sequences to determine the optimal choice of these parameters. Only one set of parameters is used for all the experiments.The configurations for Ω and Ω_{iv} are similar and equivalent to :

$$
\begin{array}{ccc}
\text{X} & \text{X} & \text{X} \\
\text{X} & \text{O} &
\end{array}
$$

where O is the current pel and X a neighbouring one.

After motion estimation process, the original image is reconstructed using the motion prediction :

$$\hat{I}(p,t) = I(p - \hat{d}, t - 1) \tag{29}$$

where $\hat{d}$ is the estimated displacement of the pel p. Motion estimation and reconstruction quality results concerning the "RUBIX" and "INTERVIEW" sequences are given by the table above in terms of percentage of uncompensated pels and compensated ones and the Mean Iteration Number of Compensated Pels ($MINCP$). The Root Mean Squared reconstruction Error values ($RMSRE$) are also given for the reconstruction error image. We clearly observe that the performances of the combined Wiener-Walker algorithm are quite satisfiying. This new proposed algorithm appears more efficient both for the improvement of the total number of compensated pixels and the motion estimation accuracy which is improved nearly about 5 to 9db if we consider the RMSRE as such an accuracy measurement.

This motion algorithm , described in this section for monoresolution image sequences , provides us with a solid statement to extend it to the multiresolution case. . The concept of neighbourhood system, naturally introduced by the Wiener-based estimator formalism, can be easily extended to others fields of image observations. In the subsequent section, we propose to test this approach in the context of subband decomposition.

4 MULTIRESOLUTION MOTION ESTIMATION

4.1 Problem statement

Several previous studies [26][27] have illustrated the efficiency of multigrid strategy within feature estimation processes. To extend motion estimation techniques to hierarchical data structures, such as pyramid transforms, we can also design multigrid motion estimator which should enable cooperation between pyramid levels. Three main problems have to be solved :

1. *The choice of a motion estimation framework.* Obviously, we will concentrate our investigation on the previously introduced modified-Wiener approach

2. *The choice of a pyramid transform.* Subsequently to the comparative results presented in Section 2 , we opt for wavelet decompositions

3. *The design of primitive propagation strategies* inside the pyramid representation : coarse-to-fine or somewhat oscillating propagation scheme.

In our case, the primitives will be the motion field; some expansion and interpolation operators are defined to control the information exchange between the pyramid levels. [28],[13] have already experimented these techniques and shown the efficiency of multigrid motion estimation schemes ; the two main motivations to use such a multigrid approach are to improve the estimation accuracy and to decrease the computational complexity .

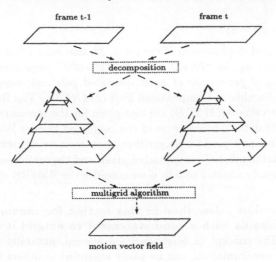

Figure 7: Multigrid approach of motion estimation on lowpass pyramids.

4.2   Multigrid Walker-Rao algorithm on wavelet pyramids

The multigrid strategy associated to a usual motion estimation algorithm as Walker-Rao method is now extended to wavelet pyramids. This choice is very attractive following the point of view that computation complexity could be decreased if we take advantages of the compactness of available data. However, the particular properties of orientation and frequency partition of the wavelet transform make the associated multigrid strategy different and more sophisticated than the usual strategies [28]. Essentially, the typical non-linear processing included within the pyramidal decomposition itself - in particular, the decimation process - eliminates the spatial-invariance properties of any image feature fields extracted on the subsampled observations

Nethertheless, assuming that we have four motion fields $\hat{d}(HH_1), \hat{d}(HB_1), \hat{d}(BH_1)$ and $\hat{d}(BB_1)$ corresponding to the four subbands of a one-level wavelet pyramid sequence, there will be two ways to reconstruct the original image :

a) by merging the four motion fields to get one field which predicts correctly the original image.

b) by using the wavelet reconstruction algorithm with the predicted subbands $H\hat{H}_1, H\hat{B}_1, B\hat{H}_1$ and $B\hat{B}_1$.

In the case (a) the main difficulty is to design an efficient merging method; some heuristic procedures such as adaptive and linear combinations have been tested [29], [12] but failed, essentially because all these merging operators have to be considered as post-processings quite independent to the motion estimation itself. All the different motion fields are computed separately and so, the optimization schemes which are duplicated in the four independent motion estimators operate separately and have absolutely no theoritical reasons to converge to the same local optimal values . Moreover, with no doubt, these four subsampled motion vector fields have absolutely no relationships with the whole motion field if it would be directly calculated on the original resolution.

In case (b) reconstruction errors of pyramid levels only affect the global reconstruction quality. The regularity and frequential bandpass properties of the synthesis wavelet filter banks could emphasize or reduce local reconstruction errors using motion estimation .

Secondly, we have to solve how to estimate the four motion fields. There are two possibilities :

c) we can use a simple monoresolution algorithm in each subband. In this case, one-level wavelet pyramids are sufficient (see figure 8).

d) we can introduce multigrid estimation, as in lowpass pyramids, (7) along each oriented N-level wavelet pyramid to improve the motion field quality. This strategy is shown in figure 9.

If $\hat{d}(BB_1)$ is estimated directly from BB_1, the amount of involved data exceeds the available one in the wavelet pyramids, else if $\hat{d}(BB_1)$ is derived from the top pyramid level BB_N then it involves some merging procedure.

Finally, one can say that a Walker-Rao multigrid algorithm on wavelet pyramid suffers from the unavailability of efficient merging operators. Moreover, the data compression requirements make the problem more complex. Because of all these desagrements, we retire this strategy in favour of a method which includes implicitely merging operators. The multiconstraint framework described in the next section provides us with such an efficient algorithmic tool.

5   MULTICONSTRAINT ALGORITHM

In the context of image sequence codecs where real-time constraints are very strong, the critical algorithmic step in terms of computation amount is essentially the motion estimation process. So in this method, the basic assumption, when several frequency-oriented

246

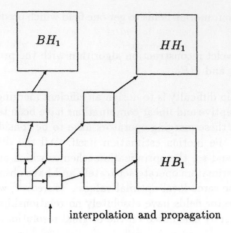

Figure 8: 1st multigrid strategy on 3-level wavelet pyramids.

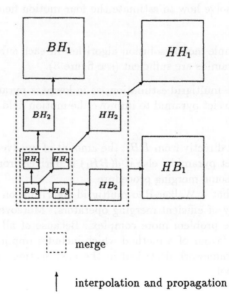

Figure 9: 2nd multigrid strategy on 3-level wavelet pyramids.

subbands are used is that only one motion vector field can be estimated and validated at all decomposition levels. We assume indeed that spatio-frequency localization properties are compatible with correlated motion attributes. From an algorithmic point of view, the main difficulty is to design an efficient merging procedure to estimate the motion vector of moving pel from several frequency bands simultaneously.

The solution that we propose is to use the combined Wiener-Walker algorithm with a multiresolution neigbourhood Ω_m. As in the usual Wiener approach where a spatial coherence is assumed for motion vector field, we suppose that a frequential coherence exists and can be useful to motion estimation efficiency. This last assumption appears quite reasonable because for pixels related to frequency components of the same moving object are related to the same motion displacement vector. By this way, we essentially use the available spatio-frequency localization of pyramid transforms to solve the initial and crucial merging problem.

Some simplifications are introduced to facilitate implementation complexity and then simultaneous combination includes only the levels which have the same size grid.

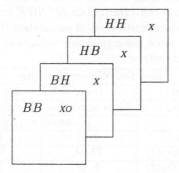

Figure 10: Multiresolution neibourhood Ω_m configuration.

So N observations will be extracted from four subbands to estimate a reliable displacement for the entire Ω_m. Assuming that we need to estimate the correct displacement of a pel p, the set Ω_m will be defined as:

$$\Omega_m = \{BB_1(p-1), BB_1(p), HH_1(p), BH_1(p), HB_1(p)\}$$

and all the tests performed inside of the estimation process will be referenced to $BB_1(p)$. The estimated displacement is then used to predict all $(\Omega_m - \{BB_1(p-1)\})$ because these four points are related to frequency components of the same spatial location and, may be, to a unique moving object.

The essential properties of this multiconstraint motion estimation can be summarized as follows:

- To make easy the comparison between the combined Wiener-Walker algorithm (spatial neighbourhood of 5 points) and the multiconstraint algorithm, we choose a multiresolution neighbourhood Ω_m with 5 points as shown in figure 10.

- The multiconstraint algorithm decreases the computation complexity in comparison to the combined Wiener-Walker applied to an original image because it works at $\frac{S_0}{4}$ image size .

- The resulted motion field is used to predict the four subbands $B\hat{B}_1, H\hat{H}_1, B\hat{H}_1, H\hat{B}_1$ and the original image is reconstructed by the wavelet reconstruction algorithm.

- The estimation process is concentrated at the lowpass frequencies (BB_1) since it represents the context of the image and contains much more information (see entropy values in section 2.2).

Comparative results between a monoresolution approach and a multiresolution multiconstraint one are given by the table below; these results only concern the low frequency band BB_1. The main important concluding remark is to observe that, for quite similar (even better) reconstruction quality ($RMSRE$ criterion) and motion compensation efficiency (% uncompensated pel number), the use of multiresolution neighbourhood in comparison to a spatial one enables a faster convergence (see $MINCP$ values) of the recursive motion estimator . Moreover, if some less critical set of parameters ($imax, T ...$) would be tested, we should observe an even more distinguishable behaviour of these two approaches in favour of the multiconstraint method.

Modified-Wiener algorithm		
Parameters	$imax = 5, T = 2, \mu = 5$	
Algorithm	Monoresolution	Multiconstraint
Sequence	INTERVIEW (BB_1)	
%uncomp.	4.80	4.48
%comp.(Test1)	73.04	72.23
%comp.(Test2)	0.75	0.75
%comp.(Test3)	21.40	22.55
%comp.(Test4)	8.34	5.36
$MINCP$	3.77	3.08
$RMSRE(BB_1)$	2.41	2.28
$SNR(db)$	40.5	41.0

Secondly, we also want to compare respectively the original image reconstruction quality between a monoresolution motion compensation approach (see table of section 3) and a multiconstraint algorithm (see table below). In both cases, a high quality of reconstruction is achieved with no perceptible reconstruction artefacts even if the $RMSE$ value decreases in the case of the multiconstraint method. Moreover, the efficiency of motion compensation is quite similar (see %uncomp. : i.e for "INTERVIEW" 3.81 in the monoresolution case, 4.48 in the multiconstraint case).and the convergence speed is improved using the multiconstraint approach (see $MINCP$ values which are respectively 3.08 (multiconstraint) and 3.55 (monoresolution) **The main advantage of the multiconstraint algorithm is that it works only on a $\frac{S_0}{4}$ image size.**

This multiconstraint algorithm could be easily extended to the lower pyramid levels by the design of any multigrid strategy.

Multiconstraint algorithm		
Parameters	$imax = 5, T = 2, \mu = 5$	
Sequence	RUBIX	INTERVIEW
%uncomp.	3.33	4.48
%comp.(Test1)	83.77	72.23
%comp.(Test2)	0.42	0.75
%comp.(Test3)	12.48	22.55
%comp.(Test4)	3.41	5.36
$MINCP$	3.50	3.08
$RMSRE(BB_1)$	1.45	2.28
$RMSRE(HH_1)$	1.82	2.08
$RMSRE(HB_1)$	1.87	3.50
$RMSRE(BH_1)$	4.31	3.64
$RMSRE$ recons.	5.00	4.86
SNR (db)	34.2	34.4

6   DISCUSSION

For television image sequence coding, pel-recursive motion techniques have been investigated using a monoresolution data structure. In this study, we have implemented several multiresolution structures to evaluate them using the same criteria. As concluding remarkes, we could say that orthogonal transforms (especially wavelet pyramids pyramids) provide us with a more compact representation in a spatial point of view. However, these methods need efficient merging operators to estimate a unique feature (motion in our case) field along the different frequential subbands and to reduce the computational complexity. A combined Walker-Rao and Wiener-based approach gives a natural algorithmic framework to implement such a multiconstraint algorithm. The results on real TV sequences validate this method giving for a suitable quality of reconstruction (compatible to broadcast quality) a decreasing of amount motion estimates : only an estimation field upon size of $\frac{S_0}{4}$ (if S_0 denotes the original image size) is needed when a multiconstraint method at one pyramidal level is used ; as further researches seem to prove, we can also introduce multigrid-multiconstraint approach using several pyramidal levels given a more smoothed motion field for an estimation surface of $\frac{S_0}{3}$.

Further results [8], [12] tend to prove that, even if such a reduced amount of motion estimates is evaluated and used to design a motion compensation prediction loop in a complete coding scheme, the compression results are quite satisfying. As preliminary but nethertheless promising results, [12] shows that, using a coarse motion compensated coding scheme based on

- the previously depicted multiconstraint motion estimation framework

- an adaptive , causal and pel-recursive motion vector predictor using these motion values

- a fixed , scalar and linear quantizer of the motion compensation errors

efficient compression rations could be achieved. As an example, for INTERVIEW sequence, a compression rate of *10* is achieved with a 38db SNR. Many further researches including some more sophisticated visual-based quantizers would significantly improve these preliminary coding results.

References

[1] VETTERLI M. Multi-dimensional sub-band coding : some theory and algorithms. *Signal Processing*, 6:97–112, 1984.

[2] WOOD J.W. and O'NEIL S. Subband coding of images. *IEEE Trans. on Acoustic Speech and Signal Processing*, 34(5):1278–1288, October 1986.

[3] BURT P.J. and ADELSON E. The laplacian pyramid as a compact image code. *IEEE Trans. on Communications*, 31:532–540, April 1983.

[4] MEYER Y. Ondelettes, fonctions splines et analyses graduées. In *Rend. Sem. Mat.*, page 45(1), Université Polytechniques de Torino, 1987.

[5] DAUBECHIES I. *Orthonormal bases of compactly supported wavelets.* Technical Report, AT&T Laboratories, USA, 1987.

[6] MALLAT S.G. A theorie for multiresolution signal decomposition: The wavelet representation. *IEEE Trans. on Pattern Analysis and Machine Intelligence*, 11(7):674–693, July 1989.

[7] BAAZIZ N. and LABIT C. Multigrid motion estimation on wavelet pyramids for image sequence coding. In *7'th Scandinavian Conference on Image Analysis*, pages 1053–1061, Aalborg, Denmark, August 1991.

[8] BAAZIZ N. *Thése de Doctorat: Approches d'estimation et de compensation de mouvement multirésolutions pour le codage de séquences d 'images.* PhD thesis, Rennes-I University, 1991.

[9] NETRAVALI A.N. and ROBBINS J.D. Motion compensated television coding-part 1. *Bell System technical Journal*, 58(3):629–668, 1979.

[10] WALKER D.R. and RAO K.R. Improved pel-recursive motion compensation. *IEEE Trans. on Communications*, 32(10):1128–1134, October 1984.

[11] BIEMOND J. , LOOIJENGA L. and BOEKEE D.E. A pel-recursive Wiener-based displacement estimation algorithm. *Signal Processing*, 13:399–412, 1987.

[12] BAAZIZ N. , LABIT C. Multigrid-multiconstraint motion estimation on wavelet pyramids for image sequence coding. *Submitted to IEEE Trans on Image Processing*, IRISA internal report:, 1992.

[13] MILLOUR C. *Contribution à la vision dynamique : une approche multi-résolutions et multi-traitements*. PhD thesis, Université de Paris-Sud, France, Mars 1989.

[14] MEER P., BAUGHER E.S. and ROSENFELD A. Frequency domain analysis and synthesis of image pyramid generating kernels. *IEEE Trans. on Pattern Analysis and Machine Intelligence*, 9(4):512–522, July 1987.

[15] SJOBERG F. *Laplacian pyramids : experiments with a multiresolution image representation*. Technical Report NA-E8569, TRITA, Department of Numerical Analysis and Computing Science, Royal Institute of Technology S-10044 Stockholm, Sweden, January 1986.

[16] BAAZIZ N. et LABIT C. *Transformations pyramidales d'images numériques*. Technical Report 526, IRISA/INRIA, Rennes, Mars 1990.

[17] ESTEBAN D. and GALAND C. Application of quadrature mirror filters to split band voice coding systems. In *International Conference on Acoustic, Speech and Signal Processing*, pages 191–195, Washington, USA, May 1977.

[18] VAIDYANATHAN P.P. Quadrature mirror filter banks, m band extensions and perfect-reconstruction techniques. *IEEE Magazine on Acoustic Speech and Signal Processing*, 4:4–20, July 1987.

[19] LE GALL D.J. and TABATABAI A. Sub-band coding of digital images using symmetric short kernels and arithmetics coding techniques. In *International Conference on Acoustic, Speech and Signal Processing*, pages 761–764, 1988.

[20] ANTONINI M., BARLAUD M., MATHIEU P. and DAUBECHIES I. Image coding using vector quantization in the wavelet transform domain. In *International Conference on Acoustic, Speech and Signal Processing*, pages 2297–2300, Albuquerque, New Mexico, April 1990.

[21] COHEN A., DAUBECHIES I. and FEAUVEAU J. C. *Biorthogonal bases of compactly supported wavelets*. Technical Report, AT&T Laboratories, USA, 1990.

[22] VETTERLI M. Wavelets and filter banks : relationships and new results. In *International Conference on Acoustic, Speech and Signal Processing*, pages 1723–1726, Albuquerque, New-Mexico, April 1990.

[23] ADELSON E., SIMONCELLI E. and HINGORANI R. Orthogonal pyramid transforms for image coding. In *Proc. of the SPIE Conf. on Visual Communications and Image Processing*, pages 50–58, 1987.

[24] SMITH M. and BARNWELL T. Exact reconstruction techniques for tree structured subband coders. *IEEE Trans. on Acoustic, Speech and Signal Processing*, 34(3):434–441, June 1986.

252

[25] FEAUVEAU J.C. *Analyse multirésolution par ondelettes non orthogonales et bancs de filtres numériques*. PhD thesis, Université Paris-Sud, Janvier 1990.

[26] GLAZER F. *in Multiresolution image processing analysis : Multilevel relaxation in low-level computer vision. editor:Rosenfeld A., Information Sciences*, Springer-Verlag, 1984. pages 312-330.

[27] TERZOPOULOS D. Image analysis using multigrid relaxation methods. *IEEE Trans. on Pattern Analysis and Machine Intelligence*, 8(2):129–139, March 1986.

[28] ENKELMANN W. Investigations of multigrid algorithms for the estimation of optical flow fields in image sequences. In *IEEE Workshop on Motion: Representation and Analysis*, Charleston, May 1986.

[29] BAAZIZ N. and LABIT C. Use of pyramid transform for motion estimation in image sequence coding. In *Picture Coding Symposium*, Cambridge, USA, March 1990.

Figure 11: Two original frames of "INTERVIEW" sequence and their temporal difference image. The Root Mean Square (RMS) of the difference image equals to 16.80.

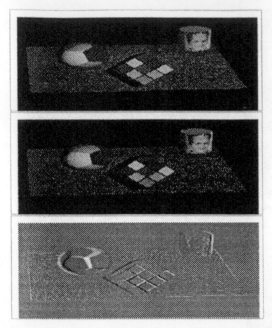

Figure 12: Two original fields of "RUBIX" sequence and their temporal difference image. The root mean square (RMS) of the difference image equals to 10.60.

Figure 13: 3- level pseudo-Gaussian and pseudo-Laplacian pyramids of "INTERVIEW" frame obtained by using 9-tap lowpass filter extracted from a QMF pair.

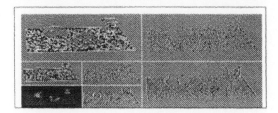

Figure 14: 2-level wavelet pyramids of "RUBIX" field generated with 4-tap wavelet filters.

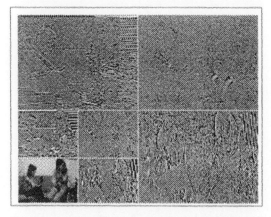

Figure 15: 2-level wavelet pyramids of "INTERVIEW" frame generated with 4-tap wavelet filters.